PUTTING SCIENCE IN ITS PLACE

science.culture

A series edited by Steven Shapin

Other science.culture series titles available:

The Scientific Revolution, by Steven Shapin

PUTTING SCIENCE IN ITS PLACE

Geographies of Scientific Knowledge

DAVID N. LIVINGSTONE

The University of Chicago Press
Chicago and London

David Livingstone is professor of geography and intellectual history at Queen's University, Belfast. A Fellow of the British Academy and the Royal Society of Arts and a member of the Royal Irish Academy, Livingstone is the author of numerous books, including *The Geographical Tradition: Episodes in the History of a Contested Enterprise.*

The University of Chicago Press, Chicago 60637
The University of Chicago Press, Ltd., London

Printed in the United States of America

12 11 10 09 08 07 06 05 04 03 1 2 3 4 5
ISBN: 0-226-48722-9 (cloth)

Library of Congress Cataloging-in-Publication Data

Livingstone, David N., 1953–
Putting science in its place : geographies of scientific knowledge / David N. Livingstone.
p. cm. — (Science.culture)
Includes bibliographical references and index.
ISBN 0-226-48722-9 (acid-free paper)
1. Science—Social aspects. 2. Science and civilization. I. Title. II. Series.
Q175.5 .L59 2003
303.48′3—dc21

2003001355

♾ The paper used in this publication meets the minimum requirements of the American National Standard for Information Sciences—Permanence of Paper for Printed Library Materials, ANSI Z39.48-1992.

For Frances

Contents

Illustrations

Preface

There is something strange about science. Scientific inquiry takes place in highly specialist sites—high-tech labs, remote field stations, museum archives, astronomical observatories. It has also been pursued in coffee shops and cathedrals, in public houses and stock farms, on ships' decks and exhibition stages. And yet the knowledge that is acquired in these places is taken to have ubiquitous qualities. Scientific findings, to put it another way, are both local and global; they are both particular and universal; they are both provincial and transcendental. To ask what role specific locations have in the making of scientific knowledge and to try to figure out how local experience is transformed into shared generalization is, I believe, to ask fundamentally geographical questions.

In large measure my interest in such questions arises from a reversal of intellectual influence in my own thinking. That journey began when I first embarked on the task of bringing the methods used by historians of scientific culture to bear on the history of geographical thought and practice. My concern was to place the development of geography, as both discourse and discipline, in the wider context of social and intellectual history. At some point along the way I began to wonder if influences might also move in the opposite direction. Could the craft competencies of the geographer, with interests in space and place, throw some light on the history of the scientific enterprise? This book is my attempt to answer that question.

Given my desire to convene conversations between geographers and historians of science, it is no surprise that I myself have benefited enormously from the stimulus and encouragement of colleagues in both communities. Accordingly, I am deeply indebted to Frank Gourley and Nuala Johnson for many helpful discussions, constructive criticism, generous reading, new leads, and fruitful suggestions. Besides the inspiration of his own work, the detailed observations of Steven Shapin on each of the chapters that follow were immensely profitable. Constructive engagement with the overall argument, as well as careful commentary on particular passages, by Adrian Johns, Robert Kohler, and Mark Monmonier was invaluable. Trevor Barnes, John Brooke, Stephen Williams, John Wilson, and Charles Withers acted as good friends should: they provided encouragement when I longed for it, criticism when I needed it, time when I asked for it, and advice when I sought it. Alas, despite their best efforts, the faults that remain in what follows are my own responsibility.

To all these I must add Susan Abrams, who rendered welcome long-distance support by telephone and e-mail. I am most grateful to her for her belief in, and enthusiasm for, this whole undertaking. Christie Henry's editorial expertise, Jennifer Howard's efficiency in responding to queries and Gill Alexander's skills in working with illustrations all did much to take the strain out of the final stages of the project. And I am much indebted to Alice Bennett for her deft and diligent copyediting. I am delighted too to acknowledge the support of the British Academy in the form of a two-year Research Readership that allowed me space to inquire into the spaces of inquiry.

A Geography of Science?

Scientific knowledge is made in a lot of different places. Does it matter where? Can the location of scientific endeavor make any difference to the conduct of science? And even more important, can it affect the content of science? In my view the answer to these questions is yes.

The suggestion that science has a geography goes against the grain. We can readily understand that there is a philosophy of science and a history of science, even a sociology of science. But the idea of a geography of science runs counter to our intuition. Science, we have long been told, is an enterprise untouched by local conditions. It is a universal undertaking, not a provincial practice. Of all the human projects devoted to getting at the truth of how things are, that venture we call science has surely been among the most assiduous in its efforts to transcend the parochial. It has been extraordinarily diligent in deploying mechanisms to lay aside prejudices and presuppositions and to guarantee objectivity by leaving the local behind. Credible knowledge, we assume, does not bear the marks of the provincial, and science that is local has something wrong with it. As one observer has put it, "It was the end for cold fusion when people decided that it only happened in Salt Lake City." Genuine science, after all, is carried on in much the same way everywhere from Boston to Beijing; experimentalists replicate each other's results in Moscow and Melbourne; conference delegates from Paris and Prague can engage in scientific conversation.

The places where science is conducted, then, seem to be of little or no consequence. Even geographers, despite their professional stake in matters of place and location, have been inclined to exempt science from the imperatives of spatial significance. To be sure, they have acknowledged from time to time that a geography of, say, astronomy could be written. But beyond such trivial circumstances as the fact that observatories are not erected in foggy valley floors or that the Pole Star is not visible in the Southern Hemisphere, there was really nothing more to say. To suggest that the methods of astronomy, or the theories astronomers devised, might be influenced by their spatial settings was little short of absurd. Of course geographers—like everyone else—readily conceded that the diffusion of scientific discoveries and technical innovations could be charted over space and time. The paths by which a new agricultural technique or medical serum spread from its point of origin, for example, could be presented in map form. But beyond such platitudinous concessions, geography seemed to have little bearing on science.

In adopting this hands-off attitude toward science, geographers have certainly not been alone. While sociologists—for long enough—were only too happy to socialize most everything from families and fiestas to rituals and religions, they drew back from looking at science in sociological terms. Whereas religion, for example, was supposed to reflect the character of the soil in which it had grown, science generated knowledge free from the imprint of the local. To be sure, certain aspects of science did seem open to sociological analysis. When scientists went off the methodological rails, allowed political prejudices to influence their research, fudged the data, read religious meaning into their findings, or came to erroneous conclusions, such "deviations" could be, and were, explained by reference to parochial factors. A sociology of what we might call "pathological science" was permissible. Or again, national and international patterns of funding and levels of state support for research were acknowledged as influencing the direction of scientific progress. But beyond either the deviant or the fiscal, there was little to say about how local circumstances might bear on the scientific enterprise. It seemed that any more comprehensive effort to situate science in the places of its making would be taken as an assault on the integrity and authenticity of scientific knowing. Indeed, the modern invention of the laboratory can be interpreted as a conscious effort to create a "placeless" place to do science, a universal site where the influence of locality is eliminated. Securing credibility and achieving objectivity required "placelessness," and the triumph of the laboratory as the site par excellence of scientific plausibility since the middle of the nineteenth century bears witness to this prevailing conviction.

This book questions such assumptions. While monumental efforts have gone into constructing "placeless places" for the pursuit of science, spaces that aspire to ubiquity, I believe there are questions of fundamental importance to be asked about *all* the spaces of scientific inquiry. What excites my interest, therefore, is the attempt to determine the significance for science of the sites where experiments are conducted, the places where knowledge is generated, the localities where investigation is carried out. The range of spatial questions we might pose is considerable. Does the space where scientific inquiry is engaged, for example, have any bearing on whether a claim is accepted or rejected? What weight is to be attached to the locations where scientific theories are encountered? In what ways has the circulation of ideas been dependent on the replication of instruments and the standardizing of methods? What strategies have been devised to acquire knowledge of things far away from direct observation? My suspicion is that along the spectrum of scales from particular sites through regional settings to national environments, the "where?" of scientific activity matters a good deal.

In anticipation of what is to follow, an illustration or two of how places matter in scientific enterprises will help to show why it is valuable to think geographically about science. During the first year of its existence in 1863, readers of Auckland's *Southern Monthly Magazine* heard the praises of Darwin's theory of evolution sung long and loud. Darwinism, they were assured, had shed new light on the settling of New Zealand by conclusively demonstrating how a "weak and ill-furnished race" inevitably had "to give way before one which is strong." Here Darwinism was welcomed because it perfectly suited the needs of New Zealand imperialists. It enabled the Maori to be portrayed in the language of barbarism and thereby provided legitimacy for land-hungry colonists longing for their extinction. At the same time, things were dramatically different in the American South. Here Darwin's theory was resisted by proponents of racial politics. Why? Because it threatened traditional beliefs about the separate creation of the different races and the idea that they had been endowed by the Creator with different capacities for cultural and intellectual excellence. For racial reasons, Darwin's theory enjoyed markedly different fortunes in Auckland and Charleston. In these two places Darwinism meant something different. In one place it supported racial ideology; in another it imperiled it.

This case could be vastly extended, as we will later see. Darwinism meant different things in Russia and Canada; it meant different things in Belfast and Edinburgh; it meant different things in workingmen's clubs and church halls. And much the same was true of Newton's mechanical philosophy, of Humboldt's global physics, and of Einstein's theory of relativity. Their accounts were understood differently in different locations and were mobilized for different cultural and scientific purposes. Scientific theory evidently does not disperse evenly across the globe from its point of origin. As it moves it is modified; as it travels it is transformed. All this demonstrates that the meaning of scientific theories is not stable; rather, it is mobile and varies from place to place. And that meaning takes shape in response to spatial forces at every scale of analysis—from the macropolitical geography of national regions to the microsocial geography of local cultures.

Space and place can be scientifically important in other ways too.

Charles Elton's theory of animal communities, for example, was born in a very specific place—on Bear Island in the Arctic during the early 1920s. And later his successor, Raymond Lindmen, developed Elton's trophic scheme through his work at another particular location, Cedar Creek Bog, Minnesota. In both cases the natural places where biotic inquiries were conducted were fundamental to the scientific knowledge generated. They were isolated sites, and their natural features made it possible to restrict variables and to carry out comprehensive measuring. They required the development of a highly specific range of what have been aptly called "practices of place." For these investigators, as for field scientists more generally, *place* was centrally implicated in everything they did. To such practitioners, the "where" of inquiry was fundamental to the practice of authentic science. Particular physical places shaped scientific theories of ecological succession, animal communities, and dune morphology.

In these ways and in many others, issues of space—location, place, site, migration, region—are at the heart of scientific endeavor. But before pursuing such cases in any further detail, it is important to reflect a little on the nature of "space" and on its role in social life more generally. Once we have grasped something of how central places are in the constitution of society, we will more easily begin to discern the inescapably spatial nature of science.

Space Matters

Human activities always take place somewhere. Where you live on the earth's surface makes a difference to the life you lead. Your location, locally and globally, has much to do with the economic, social, and cultural circumstances you find yourself in. There is an uneven geographical distribution of resources, a discernible pattern of ways of life across the face of the earth, and a distinctive spatial arrangement of the planet's physical features. Accordingly, it makes sense to speak of a regional economy, international geopolitics, a nation's cultural landscape, the social morphology of a city, or the map of world religions. These facets of human life have an obviously spatial dimension, and where an individual, a social group, a state, or a subcontinent is located in material space is therefore highly significant.

But we do not just inhabit material spaces. We also occupy a variety of abstract spaces, and we refer in spatial ways to the intellectual, social, and cultural arenas through which we move. People close together physically may be "miles apart" in terms of social distance or cultural space, living, as it were, in totally different worlds. So it is not surprising that we routinely resort to cartographic and other spatial metaphors: we hear of projects to map the human genome; some speak of theories as maps to enable us to find our way around; we are told that we each have our own mental map; we all try to chart our way through an argument or map out a course of action. In everyday speech it is common to find people wanting more personal space, feeling disoriented, or believing they are out of place. Both materially and metaphorically the spatial matters a great deal.

The social interactions we engage in from day to day are also crucially dependent on the shifting and overlapping spaces within which we transact the affairs of everyday life. Take the different arenas where we encounter other people. These include such diverse venues as the factory floor, the sports field, the dinner party, the dance floor, the office, the home—to name a very few. Each site provides repertoires of meaning that facilitate communication. The ways people behave and relate to each other in these various places can be radically different. Indeed, it has even been suggested that such conventional designations as the "normal" and the "bizarre" depend in important ways on setting: what passes as appropriate behavior in one place may be regarded as weird or grotesque in another.

Clearly, the signs and symbols that are meant to give meaning to human actions are spatially linked. For this reason, making sense of even the simplest gestures and behaviors requires an understanding of the "imaginative universe" in which the occupants of any particular locality dwell. Familiarity with the "local customs" of the boardroom or the library or the building site or the church is fundamental to sorting out the coded messages within which communication is embedded. To figure these out requires unpacking the implications and inferences that are fixed in local structures. The task is to make particular sense of particular rules in particular places.

If these sentiments are in the right neighborhood, it is plain that space is far from a neutral "container" in which social life is transacted. Space is not (to change the metaphor) simply the stage on which the real action takes place. Rather, it is itself constitutive of systems of human interaction. At every scale from the international to the domestic, we inhabit locations that at once enable and constrain routine social relations. These sites dictate what we can say and do in particular social circumstances and—just as important—what we can't. Every social space has a range of possible, permissible, and intelligible utterances and actions: things that can be said, done, and understood. These spaces of discursive exchange are the consequences of social relations, and they are important because they are not simply about agreement; they also define what kinds of disagreements are pertinent and can be expressed.

Since the positions we speak from are crucial to what can be spoken, there are intimate connections between what we might call "location and locution." Of course, locutionary acts can no more be reduced to locational circumstance than geography can be reduced to geometry. Social spaces facilitate and condition discursive space. They do not determine it. This is to say that ideas are produced in, and shaped by, settings. They must resonate with their environments or they could not find expression, secure agreement, or mobilize followers. But ideas must also be sufficiently "disarticulated" from their social environments to permit them to reshape the very settings they emerged from. Spaces both enable and constrain discourse.

The spaces of everyday social life, moreover, are not insulated from the vicissitudes of international exchange. The very opposite is the case. All of us, in one way or another, are implicated in global transactions. As the Irish poet Seamus Heaney puts it, "We are no longer just parishioners of the local." The circulation of goods and commodities, information and data, means that the local is persistently shaped and reshaped by distant influences and agents. In everything from fluctuations in the international money market to the gastronomic fantasia of almost any city where Chinese take-away, Italian pizza, and American doughnuts can all be eaten within a few square yards, the "nearby" is continuously transfigured by the "faraway." Compared with past times, of course, the pace of these processes has dramatically quickened. With such modern technologies as the telephone, the Internet, and telecommunication systems more generally, contacts between "here" and "there" are virtually instantaneous. Space has been collapsed by time. The world has, as it were, shrunk from the time when the fastest mode of transportation was the horse-drawn carriage to the era of the passenger jet. The shifting tempo and changing rhythms of this "chronogeography" have dramatically changed our world, and they point to the immense significance of commodity flows, circuits of information, and the changing relations between the present and the absent. Spaces are therefore neither static nor stable; they are mobile and mutable.

But traffic in commerce and commodity is not the only way such transformations are effected. Economic sanctions, international conflicts, and military engagements are continually reshaping the world's political map. And to that degree, at least, none of us is completely free from what has been called "the struggle over geography." That struggle, moreover, is not simply to do with armies and ammunition; it also has to do with ideas and images. How we imagine distant people and places, and how we choose to represent them to ourselves and to others, is of immense moral and political significance. Imagined geographies have real consequences.

This has long been the case. Consider Europe's rendezvous with the New World a half a millennium ago. The fabulous anthropology and mythic ecology of the Americas evoked in sixteenth-century European minds a space at once exotic and repulsive, alluring and threatening (fig. 1). The "new" realm's peoples and lands were routinely cast in the language of inferiority and barbarity, and often in the categories of those "monstrous races" long resident in European travel accounts. As a result, the Columbian encounter between the Old World and the New was at once a moral, an economic, and a sci-

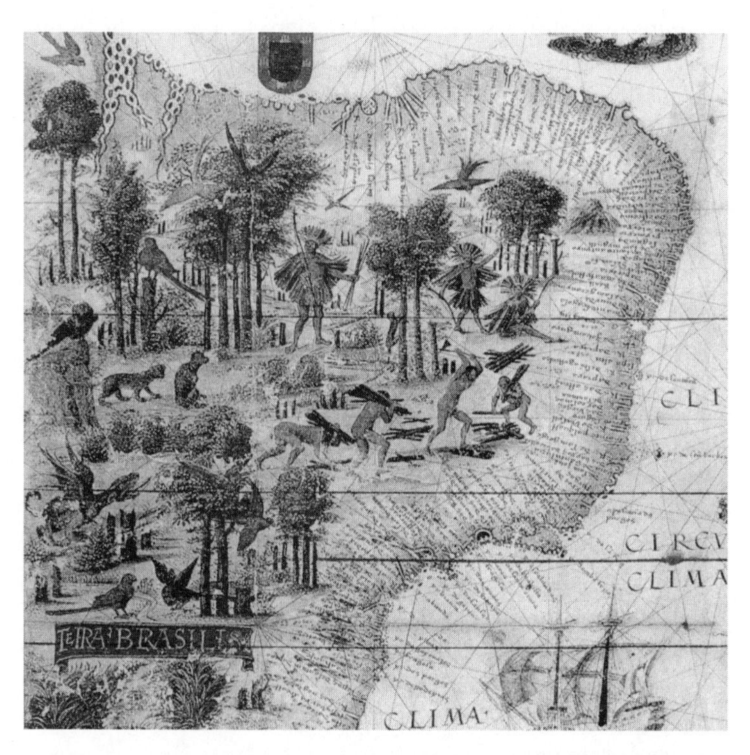

*1. This cartographic depiction of the Brazilian coast comes from the 1519 atlas of the Portuguese mapmaker Lopo Homem. The map uses conventions of representation typical of sea charts—*portolanos*—in the period. It also portrays an exotic land abundantly inhabited by native people and luxuriant plant and animal life. In the minds of Europeans, images like this conjured up imagined geographies of the New World.*

entific event. Much of the early history of this transatlantic engagement depended on Europe's geographical fantasies about the Western Hemisphere.

Nor were the Americas unique in this respect. The construction of the South Pacific as a coherent geographical entity in the eighteenth century and the designation "darkest Africa" during the Victorian era invite similar scrutiny. In both cases science and civilization conspired to bring these regional labels into currency. In the former case, the voyages of men like James Cook brought that realm cartographically, anthropologically, botanically, zoologically, and geologically within the bounds of European science. In the latter, exploration and evangelism alike contributed to the flourishing of an iconography of lightness and darkness that darkened Africa's image even as its explorers sought to flood the continent with light.

What is striking about these representations is the complicity of scientific endeavor in their propagation. The emergence of "the Orient" as a geographical region, and of the "Oriental" as a racial category, for example, was largely a product of science zealously prosecuted during and after the European Enlightenment. Through the work of geologists, engineers, anthropologists, surveyors, cartographers, and many others, Europe sought to take the measure of these newly occupied spaces. The Orient was the outcome of a cultural as well as a military intrusion, a scientific as well as a religious crusade. It was a fact-fiction fusion that set off "the East" as the alter ego of "the West." In turn, this imagined space became the locus of scholarly attention and found itself exhibited as Europe's exotic "other." On paper and canvas, in museums and exhibitions, through spectacles and snapshots, legendary geographies of the "Orient" were presented to the eyes of European witnesses (fig. 2).

All these endeavors reveal, to one degree or another, the power of place. At the world scale, the capacity to represent global regions—and thereby to construct them in human consciousness—has been fundamental to the practices of political supremacy. At the opposite end of the spectrum, very specific sites also exercise tremendous power over people. Take the hospital, the church, or the courtroom. In such spaces people are brought under the authority of medical, religious, and legal knowledge for different purposes. In these places people undergo medical diagnosis; they are told what is good and evil; they are sentenced or acquitted. In all three, they experience discipline, in one form or another, of body, mind, and spirit. For in all three there are intimate connections between the regimens of regulation that are practiced and the regimes of knowledge that govern

2. Pictures like this romanticized view of Jaffa, painted by John Carne in 1837, did a good deal to confirm European images of the Middle East as strange and mysterious, a region distant and different. Through works of this sort, "the East" was constructed in the Western imagination as an exotic geographical realm.

them. To understand the *history* of medicine, or religion, or law, then, we must necessarily grasp the *geography* of medical, religious, and legal discourses. It is critically important to pay attention to those sites that have generated learning and then wielded it in different ways. At every scale, knowledge, space, and power are tightly interwoven.

Place is essential to the *generation* of knowledge. It is no less significant in its *consumption*. Ideas and images travel from place to place as they move from person to person, from culture to culture. But migration is not the same as replication. As ideas circulate, they undergo translation and transformation because people encounter representations differently in different circumstances. If theories must be understood in the context of the period and place they emerge from, their reception must also be temporally and spatially situated. So if we are to appreciate something of how thoughts and theories, insights and imaginings, concepts and conjectures have changed the world, we need to be as attentive to how they are appropriated as to how they are made. And what is true of images and ideas in general is true of their scientific counterparts.

Geographies of Science

In what ways, then, does it make sense to speak of science as having a geography? Science is concerned with both ideas and institutions, with theories and practices, with principles and performances. All of these have spatial dimensions. Consider the laboratory as a critical site in the generation of experimental knowledge. Who manages this space? What are its boundaries? Who is allowed access? How do the findings of the laboratory's specialist space find their way out into the public arena? Attending to the microgeography of the lab—and a host of other similar spaces such as the zoo, the botanical garden, or the museum—takes us a long way toward appreciating that matters of space are fundamentally involved at every stage in the acquisition of scientific knowledge. What is known, how knowledge is obtained, and the ways warrant is secured are all intimately bound up with the venues of science.

The geography of science also calls attention to the uneven distribution of scientific information. Not everyone has had access to the deliverances of science because there are diffusion tracks along which scientific ideas and their associated gadgetry migrate. The means and patterns of circulation, understandably, have changed dramatically over the past three hundred years or so. But the movement of science has had an impact of immense proportions. Then again, it surely makes sense to reflect on whether scientific cultures themselves display any discernible political or social topography. Can certain types of scientific inquiry be correlated with certain social classes, or with those of a particular religious persuasion, or with metropolitan or provincial cultures? To what degree was the science produced in colonies colored by the cultural politics of imperial powers? Has scientific work been used to sustain the ideology of particular groups and to promote their interests over those of others?

Asking questions of this sort does not prejudge what the answers might look like, but it does suggest that our investigations will call attention to the local, regional, and national features of science. This means that science is not to be thought of as some transcendent entity that bears no trace of the parochial or contingent. It needs, rather, to be qualified by temporal and regional adjectives. At one scale of operations, science is always an ancient Chinese, a medieval Islamic, an early modern English, a Renaissance French, a Jeffersonian American, an Enlightenment Scottish thing—or some other modifying variant. While most of my discussion will rotate around science as we think of it in the West, that should not be taken to imply that these are the only practices that warrant the name science. We must work with a less fixed conception of what science is. What passes as science is contingent on time and place; it is persistently under negotiation. This becomes clear when we ask a question like, Were Plato and Aristotle engaged in the same sort of activity as, say, Newton and Boyle or Watson and Crick? To say that they were all "doing science" doesn't help much, for the meanings of the very terms we use change from period to period and from place to place. Even seemingly standard designations like "atom," "gene," and "species" have undergone transformation. And the same is true for scientific movements like "Copernicanism," "Newtonianism," and "Darwinism." Science is not some preordained entity fulfilling an a priori set of necessary and sufficient conditions for its existence. Rather, it is a human enterprise, situated in time and space.

Cultivating a geography of science will disclose how scientific knowledge bears the imprint of its location. Of course, there are constraints on what scientists can reasonably say about nature and—more important—what they can do with it. They can't just *decide* what to believe about reality. Scientists make science, but they do not do so entirely as they choose. Yet if scientific endeavor can yield true accounts about certain aspects of the world, it can do so only at particular times, in particular places, through particular procedures. This means that every aspect of science is open to geographical interrogation. Place matters in the way scientific claims come to regarded as true, in how theories are established and justified, in the means by which science exercises the power that it does in the world. There are always stories to be told of how scientific knowledge came to be made where and when it did. The appearance of universality that science enjoys, and its capacity to travel with remarkable efficiency across the surface of the earth, do not dissolve its local character. To the contrary. These triumphs are at least in part a consequence of such spatial strategies as the replication of equipment, the training of observers, the circulating of routine practices, and the standardizing of methods and measures.

The sweep of geographical questions to be asked of science, then, is remarkably extensive. In what follows I have chosen to dwell on three dominant geographical motifs—site, region, and circulation—and their consequences for science. The book, in other words, is organized spatially rather than temporally, geographically rather than historically. In this respect it departs from the conventional practice of according priority to time over space in thinking about the nature of science. Of course this is not intended to deny the profound importance of historical change, temporal progress, and cognitive shifts from one period to another. But the book's structure is designed to foreground the spatial, and this arrangement means that my argument is developed through a sequence of episodes drawn from different points in time and selected to bring into sharper relief the role of geographical modalities in scientific inquiry. It also means that there is no single, overarching formula explaining how space invariably shapes science. Indeed, to seek an account of this sort militates against the spirit of this book. For in different locations, at different times, in different circumstances, and at different scales, space had made its mark on science in different ways.

Initially we consider the sites from which scientific knowledge emerges. These range widely from the laboratory to the zoological garden, from the field to the museum, from the hospital to the public house. In each case our concern will be to ascertain the significance of these locations in the shaping of their respective scientific inquiries. And we will visit some of those hidden spaces where science is practiced in secrecy, either from fear of public protest or in clandestine explorations. The human body as a site of scientific inquiry, not only for medical research but also as itself a measuring device, will also engage our interest. Throughout, we will find scientific claims that sound universal but turn out to be situated, theories that seem transcendent but are profoundly embodied. At the same time, the plurality of scientific sites bears witness to the protean nature of science. Indeed, there is much justification for suspecting that the term "science" is an imaginary unity masking the disparate kinds of activity that trade under the label. It will be wiser, therefore, to work with the assumption that in different spaces different kinds of science are practiced.

Thereafter we turn to the larger regional scale. Here we will peruse some of the ways regional cultures, provincial politics, national styles, and such have conditioned the practices and products of scientific endeavors. There was a distinctly regional pattern to the rise of scientific Europe, and our task will be to determine why certain forms of scientific activity emerged in certain regions and at certain points in time. These reflections will confirm the salience of the geographical adjectives in "English science," "French science," and "Russian science"; they will also demonstrate why it makes good sense to think of "the Scientific Revolution" as having a geography as well as a history. By the same token the significance of more local scales, provincial and urban, will be underscored. Just why it makes sense to speak of, say, Manchester science during the Industrial Revolution or Charleston natural history in the mid-nineteenth century will become clear. The different ways novel concepts and practices have been received in various places will also attract our interest. Spaces of resistance and indifference tell us as much about the culture of science as spaces of acceptance and appropriation. By working at the regional scale, we can begin to get a sense of how local particularities shape the ways scientific theories are encountered, mobilized, or discarded.

Finally, matters of circulation will assume center stage as we ponder the significance of the movement of specimens and instruments across space and time, as we reflect on the strategies devised to glean reliable information about far-off things, and as we consider how knowledge travels from place to place. Systems of establishing trust, standardizing measurement, and disciplining observers—all key features of attempts to obliterate the cognitive distance between "here" and "there"—will necessarily command our attention. In this connection we will reflect on the use of techniques like mapping and picturing as ways of overcoming doubts about the reliability of travelers' reports and as methods of freezing time and fixing space. We will come to appreciate how what appeared to be detached findings were actually the outcome of judgment, negotiation, and regulation. The successful circulation of scientific knowledge was, as much as anything else, about settling upon strategies to stabilize knowing-at-a-distance.

I am fully aware that these items do not exhaust the scope of what I am calling the geographies of science. My arguments are built around historical examples drawn from the period between the sixteenth century and the early twentieth. Ancient and medieval science, as well as twenty-first-century science, have their own geographical narratives that need to be told. My aim is simply to provide a suite of illustrative cases, not a comprehensive survey, of how geography matters in scientific inquiry. And my hope is that these deliberations will catch the imagination of some readers who will be encouraged to venture into the terrae incognitae of scientific culture and continue the task of surveying science's hitherto unexplored spaces.

Site

VENUES OF SCIENCE

The range of sites within which science has been practiced, in which meaning has been made and remade, and from which scientific knowledge spreads is vast. We can begin to catch something of this diversity if we conjure up a mental picture of some of the disparate places where science is conducted. When we do we are impressed with the vastly different *atmospheres* they exude. The claustrophobic darkness of the alchemist's workshop with its roaring furnace and smelly stills (fig. 3) stands in marked contrast to the clinical brightness and flickering screens of the modern medical technology laboratory. The wide-open, airy spaces of the field (fig. 4) contrast sharply with the fusty alcoves of the archive and stuffed displays of the museum (fig. 5). The controlled exhibits of the botanical and zoological gardens are very different from the diagnostic spaces of the hospital or the asylum. Even to express things this way, of course, is to run the risk of caricature. Laboratories, gardens, museums, observatories, hospitals, and so on all come in a wide variety of shapes, sizes, and configurations. But these stereotypes do have sufficient imaginative currency to convey something of the range of *sensory* experiences that

3. "The Alchemist in His Laboratory," painted by David Teniers the Younger about 1649, conjures up the atmosphere of an alchemist's workroom. The scattered texts, the master using the bellows, various assistants in the background, and a variety of vials and other pieces of equipment convey the sense of a cluttered, claustrophobic space.

such sites induce with their different sights, sounds, and smells. Each constitutes a different suite of optical, acoustic, and olfactory spaces.

Scientific practice is influenced by these spatial settings in a number of ways. For a start, the disposition of equipment and other accoutrements regulates human behavior in one way or another. Frequently the site is constructed so as to restrain or promote certain interactions; in some cases entry is carefully controlled by formal or informal mechanisms of boundary maintenance. It is also within these spaces that students are socialized into their respective scientific communities. Here they learn the questions to be asked, the appropriate methods of tackling problems, and the accepted codes of interpretation. Here decisions are settled about what passes as scientific knowledge, how it should be acquired, and the means by which

4. A photogravure by Henrik Grönvold of Joseph Whitaker's encampment in the Tunisian Sahara, taken from his 1905 ornithological history of the regency of Tunis. Climatically and socially, Whitaker found this an ideal site of scientific inquiry.

claims are warranted. In these venues practitioners absorb the core values, convictions, and conventions of their tradition of inquiry. To this extent, science is always local. Whether it is a John Dee conjuring angels in his domestic studio, an Isaac Newton conducting light experiments in a darkened room in Trinity College Cambridge, an Alfred Russel Wallace mapping plant and animal distributions in Borneo, or a Josef Mengele carrying out experiments in racial hygiene at Auschwitz, the site-specific conditions of knowledge making are immensely different. So too are the ways the knowledge accumulated moves out from its site of origin into the public sphere.

Various questions, then, plausibly arise from the obviously variegated geography of the spheres of science. How, for example, is the circuit of knowledge effected, from the domain of acquisition to common currency? How does knowledge move from the particularities of its site of production to communal exchange? If, as we might suspect, specific spaces of science were not homogeneous, then how were they internally structured? Just who was permitted access to those

5. An etching of the early seventeenth-century museum of the apothecary Francesco Calzolari. The sheer abundance of material in this collection, crammed full of exotic objects, excited in viewers a wondrous sense of latent therapeutic potential.

privileged places of knowledge generation? Did the line separating "insiders" from "outsiders" map onto any other contours—say, of gender, class, status, ethnic group, or professional standing? And how was the work divided up among those who *could* cross the threshold into the knowledge-making territory? Is any significance to be attached to the locations chosen as sites for scientific pursuits? What kind of relationships pertained between the private and public within those spaces, and indeed between the specialist space of the investigator and the outer world of intellectual commerce? Such questions suggest that the spaces of science are far from incidental to the enterprise and that there are both physical and intellectual geographies of knowledge production to be uncovered.

The array of knowledge-producing sites is immense. We will thus approach them through a rather rudimentary taxonomy. This schema is only suggestive, and there are many cases of significant overlap. Our concern is simply to capture something of the forces animating scientific inquiry in these arenas.

Houses of Experiment

We have become accustomed to the idea that scientific endeavor takes place in specialized locations like the laboratory. In part this has to do with the equipment scientists need to carry out their activities. Telescopes, microscopes, pumps, retorts, test tubes, and precision instruments of all kinds need to be housed. But the placing of scientific inquiry in designated spaces cannot be reduced simply to the requirements of instrument management. There is a history here of far wider dimensions. And one way to begin thinking about the spaces of experiment is to briefly glance at the prehistory of the laboratory.

A long-standing tradition in the West was the idea that retiring from society was a precondition for securing knowledge that was of universal value. Prophets and seers withdrew into solitude and returned with insights devoid of parochial particulars. Ironically, to acquire knowledge that was true *everywhere,* the seer had to go *somewhere* to find wisdom that bore the marks of *nowhere.* Such sentiments arose from the conviction that to be authentic, the sage must stand outside the normal confines of society. It was precisely this kind of solitude that the monastic life sought to provide. But the monastery and the hermitage, not to mention the wilderness and mountaintop—all classical sites of medieval spiritual knowledge—were not suited to experimental pursuits. The ideal of solitude remained, but a new space had to be carved out to accommodate it. What, then, was the route from the monastery to the laboratory as a site of knowledge production?

To get some sense of how that new kind of space—laboratory space—began to be hewn out of existing spatial arrangements, it will be useful to pause for a moment at a house in Mortlake on the banks of the river Thames. It is the home of John Dee, Elizabethan England's most celebrated natural philosopher. Despite initial impressions, this is no ordinary residence of the gentry. Strange sounds and foul smells emanate from certain regions of the dwelling. For in the Dee household we witness an early move in the relocation of knowledge generation into the domestic scene. Rooms were dedicated to various alchemical appliances and occult practices because the master of the house needed to slice a private workspace out of an otherwise public household. Securing such a hermetic retreat within the home, of course, cannot be understood in isolation from the more general social history of the house. The unpartitioned and rather public space of the large medieval dwelling had by now progressively given way to compartmentalization; private quarters provided retreat and solitude, especially for the merchant and "professional" classes. Among the domestic innovations of the sixteenth century were back stairs and basement rooms built to serve the needs of the household. Such arrangements provided conditions into which the spatial requirements of the natural philosopher could be inserted. Since the house is an architectural expression of social structure, the experimental workshop fashioned out of domestic quarters represents an important step in the segmentation of world and of self.

Getting hold of a piece of space for alchemical experiment in his own home was a tricky business for John Dee. For one thing, it created tensions between him and his wife, Jane, at a time when the domestic roles of husband and wife were in transition. Simply put, John's experimental life got in the way of Jane's management of the household. And then the costs of distilling equipment, for example, put a strain on the family purse. Besides all this, the large number of servants in the dwelling, together with the various assistants Dee employed over the years, made finding any privacy difficult. And solitude was understood, as we have seen, to be an essential prerequisite for intellectual projects. So here, right on the cusp of the emergence of what has been called "laboratory life," Dee was embroiled in a series of negotiations between the call of the private and the demands of various publics, family and otherwise. Even Jane was forbidden to enter the room where John engaged in his astral conversations; that was a sanctuary where angelic forces transacted the business of natural philosophy with her husband. Finding himself stranded between the library and the laboratory, Dee represents a key moment in the early construction of experimental space.

But it was not just the seemingly furtive crafts of the occult sciences—frequently closeted in basements—that were secreted within inner chambers. A host of key players in the emergence of English science in the mid- to late seventeenth century had laboratories either in their own homes or in the homes of gentleman patrons. Think of the circumstances at Robert Boyle's home in Pall Mall in London, where he spent the last twenty or more years of his life in the home of his sister Katherine, Lady Ranelagh. Here, it seems, the laboratory was again in the basement; but it did have its own direct access from the street. These arrangements were significant, because while solitude was important to Boyle, he and his associates at the Royal Society insisted that scientific knowledge was in some fundamental sense a public matter. So while Boyle lamented over disturbances, he still needed to accommodate the new science's strictures on the public attestation of natural knowledge. In order to achieve the status of "knowledge," claims had to be produced in the right place and had to be validated by the right public. *Where* science was conducted—in what physical and social space—was thus a crucial ingredient in establishing whether an assertion was warranted. So Boyle's experimental quarters had to be both private and public space at the same time.

Of course the new experimental arenas that surfaced in the period, not only in Boyle's home but at the Royal Society and elsewhere, were far from public in today's sense. To be sure, "gentlemen" were permitted access, according to the social norms of the day. But most important was what has been called "the experimental public"—those whose presence was essential to the confirmation of empirical findings. Occupying the laboratory's physical space was one thing; occupying its discursive space was quite another. This means that the laboratory's *social* space was differentiated in a number of ways. It had, so to speak, its own cultural topography. On the one hand, there was a gulf between figures like Boyle, who had the authority to deliver knowledge, and the numerous attendants who worked the equipment and operated the instruments. The latter had craft competence, but they lacked the social standing to make scientific knowledge. Here was what we might call an epistemological chasm dividing the scholar from the mechanic. They occupied different social spaces. And in so doing they gave spatial expression to a suite of dualisms running the length and breadth of English society in the seventeenth and eighteenth centuries—philosopher and artisan, head and hand, mind and brain, soul and body. On the other hand, casual callers inhabited a different knowledge space from those socially and cognitively sanctioned to ensure experimental reliability. Here were boundaries that, though unmarked in physical space, were prominently displayed in the laboratory's mental cartography.

The whole issue of the public warranting of knowledge raises yet another matter of spatial significance for science. Because an experiment "worked" in the private recesses of the scientist's workplace was not sufficient to establish its claims as genuine knowledge. To secure *that* level of cognitive standing, it had to receive the approval of the relevant experimental public. A gulf thus opens up between what has been called the "trying" of an experiment and the "showing" of an experiment. Only when the journey from private to public space had been successfully concluded could a scientific claim enjoy the privilege of knowledge status. Through public demonstration, private speculation achieved open confirmation. The shift from "trying" to "showing," from delving to demonstrating, we might say, is a spatial manifestation of the move from the context of scientific discovery to the context of justification.

Stabilizing experimental claims, however, was often not just a matter of disclosing them; it was frequently necessary to dramatize them. This was as true of the spectacles that Michael Faraday put on for his Victorian audiences at the Royal Institution as it continues to be of the nuclear power industry. As for the former, Faraday's famous Friday evening performances during the 1830s and 1840s, in the presence of a highly controlled guest list, were presented in such a "natural" manner that any sense of artistry was erased. His hard work behind the scenes was as effective in the way he presented "nature" as were the elocution lessons he had taken to improve the way he presented himself. In the latter, the demonstrations are so effective precisely because the smooth public performance obscures the untidiness behind the scenes; the vagaries of nature are caged, as one observer puts it, "in thick walls of faultless display." Here the theatrical dimensions of experimental space are clearly exposed.

Demonstrations, however, have long trailed their own clouds of cynicism. Public experimentation could easily be charged with using illusory techniques to deceive the eye. So during the eighteenth century, natural philosophers felt the need to find ways of putting distance between themselves and those plebeian "mechanicals" more intent on dramatic entertainment. The carefully constructed experimental arena, requiring sophisticated instrumental choreography, could too easily resemble the trickery of the mountebank. All this meant that experimental display inhabited a space poised between conjuring tricks and scholarly authority, between the theater and the academy. Nevertheless, what experimental demonstration succeeded in establishing, certainly over the long haul, was a way of knowing that required hands-on experience irreducible to conventional numerical or linguistic signs. That this outcome was the product of long-drawn-out negotiations is nicely disclosed in the origins of the modern university physical laboratory, where space had to be appropriated to provide instruction and demonstration to students, on the one hand, and research facilities for teachers, on the other.

The founding of the Cavendish Laboratory in Cambridge in the 1870s, and of its Scottish predecessor, William Thomson's experimental lab in Glasgow a decade and more earlier, nicely illustrate such maneuvers. The founding of the Cavendish required the acquisition of a species of intellectual and material space hitherto alien to the university's academic ethos, for mid-Victorian Cambridge was the stronghold of Anglicanism and mathematics. The workshop was terra incognita to the university establishment. With its savor of the merely technical, moreover, it threatened the moral fabric of the old order by transgressing lines of social demarcation. Such was the environment into which proposals for a new physical laboratory were launched as a key feature of the move to bring experimental physics into the English academy. Territorial acquisition, it seems, is as fundamental to educational crusades as it always has been to military campaigns. To grasp the factors involved in this reconfiguration of the geography of Cambridge science, it is illuminating to glance at the apologia for the new space made by James Clerk Maxwell, the nineteenth-century Scottish physicist and student of electromagnetism.

Maxwell knew only too well that the values of the factory workshop were alien to the dominant university ethos of his time, and that he needed to find some way of domesticating the world of the laboratory to the prevailing culture of scholarship. In fact, as a Scotsman he was particularly well suited to the project of mediating between the scientific reformists pushing for an electromagnetic and thermodynamic laboratory and defenders of the traditional mathematics curriculum. For, as was typical of late Enlightenment Scottish intellectuals, he retained a strongly metaphysical cast of mind, and he applied it to the disputed connections between algebra and geometry. This enabled him to urge that providing facilities to ensure precise computational standards was analogous to the Anglican God's work of calculation and measurement. Thereby he could forge a strategic alliance between God and mammon, between philosophy and the factory. Not surprisingly, displayed over the Gothic entrance to the Cavendish, suitably decorated with the family coat of arms, were words from the book of Psalms. The new physical laboratory was a spatial and symbolic incursion into the university's scholarly domain. The chapel and the study now had to make room for the laboratory. Work carried out here on electrical technologies would transform the social polity. Even as the physics laboratory recast the academy, the telegraph remade global geography.

Maxwell's laboratory labors, however, were not without precedent. Entirely appropriately for a Scotsman, he had looked for inspiration to William Thomson—Lord Kelvin—in Glasgow. For here Thomson had cemented the very links that Maxwell yearned for between the culture of the classroom and the craft of the foundry. Glasgow College—one of Scotland's five medieval university colleges—was particularly well positioned to move beyond what has been called the "monkish" values of Oxbridge. After all, among its celebrated achievements the city boasted James Watt's steam engine and Adam Smith's political economy, both symbolic of its progressivist inclinations. As Thomson himself remarked, Glasgow was especially well served with suppliers of the apparatus of the new industrial order. Add to these the visual drama that attended his spectacular demonstrations in electricity and magnetism, and we find the ingredients that enabled Thomson to wrest the intellectual initiative from those wedded to the old regime long enshrined in the college's hallowed halls. And so in Glasgow, as later in Cambridge, the physical laboratory emerged as the spatial articulation of a new cultural order (fig. 6).

Having looked in on a set of rather different "houses of experiment," it is clear that we have witnessed a variety of laboratory microgeographies. Concurrently we have seen that laboratory space has conveyed a range of meanings. There have been occasions when it assumed the role of theater; as knowledge moved from its point of origin to public disclosure it frequently had to be dramatized in order to be stabilized. The space of experiment was also theatrical in that this is where various stagings of nature took place; in the microworld of the lab, aspects of the world were manipulated, controlled, and reconstructed courtesy of the available technology and the local experimenter's know-how. Indeed, it was only by operating material apparatus in the laboratory that such invisible entities as lines of magnetic force could be made manifest. At the same time the laboratory's very construction was routinely seen as a decisive cognitive move in the campaign to establish new ways of knowing. The laboratory was thus an emblematic space replete with cultural meaning, though as a site of knowledge it could function only in the presence of the geo-

6. The inner quadrangle of Glasgow College: the natural philosophy classroom is on the first floor between the turrets. William Thomson's establishment of the physical laboratory, with its strongly industrial associations, represented a radical departure in Scottish university education. Securing this laboratory space was a key element in Thomson's program of educational reform.

graphically privileged who were permitted to cross the threshold. Their role was critical. For only if they enjoyed the trust of those *outside* could they warrant the credibility of the claims made *inside* the laboratory's walls. They were the vehicle by which experimental knowledge was "disembedded"—extracted from its place of origin—and transferred to a wider public.

Cabinets of Accumulation

The experimental laboratory, of course, is not the only site where scientific endeavor has taken place. Running alongside, and indeed predating, the laboratory were spaces of accumulation such as the museum and the archive, where specimens and samples were collected and organized according to the prevailing norms. In these chambers the aim—at least in part—was less to manipulate the natural world by experiment than to arrange it through classification. Whereas the drama of the laboratory lay in staging demonstrations, the museum's theatricality is expressed by amassing, ordering, categorizing, and displaying exhibits of all kinds.

The origins of museum culture can be traced back to what were known as "cabinets of curiosities"—early collectors' closets into which gentlemen of the sixteenth century packed curios of all kinds. The more rare in occurrence or exotic in appearance or distant in origin an object happened to be, the more likely it would end up in a "world of wonders" housed in some secluded antechamber. This habit of collecting, frequently nurtured as a noble pastime, was actively cultivated as the insignia of a civilized household and, it has been said, provides "a window into a private psyche" by disclosing the whims of the collection's proprietor. As instances of conspicuous consumption, collectible objects served to confirm social standing.

At the same time, the acquisitive impulse delivered, in the form of the museum, a key site for pursuing a different kind of scientific knowing. The dazzling variety of the natural order, with its profusion of particularity, which the museum accumulated, classified, and re-presented, did much to feed the nascent scientific craving for facts, more facts, and yet more facts. In contrast to Albertus Magnus, the medieval Scholastic philosopher who had insisted in the mid-thirteenth century that "there can be no philosophy of particulars," the English essayist, statesman, and philosopher Francis Bacon, in his *Novum Organum* of 1620, called for "particular natural history" and the accumulation of "Singular Instances." For Bacon these very things were crucial to overthrowing a priori thinking, impromptu generalization, and the syllogistic reasoning so beloved of contemporary natural philosophy. To be sure, wonder might be nothing but open-mouthed gawking or vain admiration, pernicious astonishment or reverential awe, but when harnessed to curiosity it could do scientific work. Thereby collecting became established as a valuable and valid way of knowing. Concurrently, the idea of wonder came to reside both in material objects and in the human response those artifacts excited. All this meant that a seventeenth-century collection might exhibit, side by side, dwarf species, chameleons, wampum belts, mathematical instruments, Turkish knives, Oriental footwear, stuffed dodoes, and medals of famous men. Such seemingly bizarre juxtapositions were cataloged according to the conditions of their acquisition, their philological associations, and their historical circumstances. In this way the museum, as a site of accumulation, played a vital role in the mushrooming of those sciences concerned with ordering and arranging the specimens of natural and civic history.

Originating in the *studio,* by the end of the seventeenth century the museum had become a *galleria.* And this shifting internal geography had important ramifications for the kind of place it turned out to be. As a setting for scientific inquiry and human interaction, the museum was—both socially and acoustically—a synthetic space. It mediated between private and public domains, yet it was, as one scholar puts it, "located between silence and sound." In its early days, as the stillness of the study yielded to the murmur of the gallery, the museum provided a setting for courtly—and almost always manly—civility in which the virtues of scholarly conversation could be engaged (fig. 7). As it renegotiated the relationships between intimacy and so-

7. The late sixteenth-century museum of Ferrante Imperato, in the family palace. In contrast to the solitariness of the study, this illustration reveals the museum as a space of male conversation as well as a site of natural curiosities.

ciability, the domestic and the public, museum space at once socialized privacy and cloistered civility. Indeed, some felt that when the Ashmolean Museum at Oxford allowed access to the public, including women, in the late 1680s, the polite boundaries of the community of learning had been grossly breached.

Because the gallery was no longer a static site of contemplation but had become a mobile space through which patrons passed, it signaled a move away from contemplation toward the active life as the road to genuine knowledge. Bodily movement, intellectual exchange, and ordered display were all integral parts of a domain whose very existence was dependent on a never-ending ebb and flow of commodities, information, and conversation. But as articles streamed in from near and far, they were reassembled, positioned, and displayed in the way the curator believed was most appropriate. So even while museums exhibited real-world objects, they refashioned reality through classification, location, and genealogy. Museums have thus always been sites of interpretative practice in which the spatial allocation of items fundamentally reconfigures the world of nature.

At no time, perhaps, was the obsession with amassing and arranging global data more feverishly nursed than at the height of Victorian Britain's overseas imperial adventure. From institutions like the Royal Geographical Society, the Royal Asiatic Society, the Royal Society, and most particularly the British Museum, the acquisitive tentacles of empire snaked their way around the globe. Yet the fact-fascination that characterized such spaces ultimately reduced universal geography to the cabinet-sized exhibit and file-sized archive. In turn these became the way administrators, bureaucrats, and the general public encountered the "collective improvisation" that was the British Empire. By accumulating, reorganizing, and reproducing information from the remotest corners of the earth, the Victorian archive played its part in shaping worldwide geopolitical relations. In one way or another, the data-hungry museum did much to fulfill the surveillance needs of colonial management.

Because it is a vehicle for expressing knowledge claims, the museum's spatiality has often been an arena of struggle. In Charles Willson Peale's museum in Philadelphia, which first opened its doors to the public in 1786, tensions arose over classifying North American specimens according to Carolus Linnaeus's fundamentally European system. The new republic and the Old World were locked in cultural combat over the relative excellence of the two continents' floral and faunal specimens. Yet Peale felt constrained to adopt the Linnaean system even while expressing a preference for native American nomenclature. His museum thus further advanced the domain over which the European taxonomists held sway. By the same token, Peale's remarkable undertaking did much to constitute the museum as a vital educational tool in the life of the new democracy. Ordinary people like farmers and merchants, he was sure, could benefit from the commercial possibilities of natural history.

Again, when Alpheus Hyatt was hired as the permanent curator of the public museum in Boston in 1870, he immediately set about using the collection to illustrate the development of species. Thereby he dramatically departed from the creationism of his teacher, Louis Agassiz, who once observed that the "great object of our museums should be to exhibit the whole animal kingdom as a manifestation of the Supreme Intellect." Hyatt, by contrast, regrouped exhibits into a set of categories that revealed the development of species—mineralogy, botany, paleontology, zoology, geography, and anthropology. Such moves were far from inconsequential, because the museum had by now become a significant teaching venue within American colleges. Agassiz's own Museum of Comparative Zoology at Harvard, for instance, superseded the classroom as his key site of instruction. And later in the 1930s, at the American Museum of Natural History in New York, the differing views of William King Gregory and Henry Fairfield Osborn on the evolution of primates found expression in their respective exhibition halls. Gregory's "Hall of the Natural History of Man" stressed the evolutionary continuity between the different human races, whereas Osborn's "Hall of the Age of Man" sought to undermine the theory of ape ancestry, to stress parallel development, and to portray the different human races as discrete "species." The displays mounted on the second and fourth floors of the museum thus articulated the different social, political, and religious convictions of the two scientists. In ways like this, the museum voiced the values of its curators and disclosed their mental geographies.

The museum, then, can be considered a map of its curators' claims to knowledge. Richard Owen, the celebrated English comparative anatomist and first director of the Natural History Museum in London, expressly couched his tale of the genesis of the building in the language of organic progression. The structure itself, he assured his hearers in 1881, displayed developmental advances in morphology, and the arrangement of its specimens reflected his own conception of natural history. The same was true of the Museum of Practical Geology in London's Jermyn Street, where the geological displays in the early Victorian period were arranged so as to establish and stabilize the version of stratigraphy championed by Sir Roderick Murchison. Precisely because items were allocated to their "proper places," the layout of the galleries was itself a map of geological knowledge.

Much the same was true of anthropology. The regulation of museum space in late nineteenth-century America conjugated the differences between the anthropology of Franz Boas and of Otis T. Mason and his mentor John Wesley Powell. Whereas the latter two employed an evolutionary narrative to account for—and to display—certain ethnographic inventions, Boas urged the virtues of exhibition by tribal group. To Mason and Powell the very purpose of the museum was to reveal progress— of anthropology, of science, of human culture. For Boas, ever impatient with taxonomic systems, schemes of unidirectional evolution, and what we might call "object fetishism," the goal was to confirm the relativity and diversity of human civilization. How space was managed declared differences between evolutionary and ethnic modes of anthropological understanding, between temporal and territorial ways of thinking. The profound contrast between anthropological leaders on the very nature of their projects was literally laid out in the layout of the exhibits. The microgeography of museum pathways disclosed different ways of telling the story of the human species.

Such stories, moreover, could have wider ideological implications. In his museum of anthropology in Victorian Oxford, Augustus Pitt-Rivers used his displays—whether of weapons, tools, pottery, or religious paraphernalia—to emphasize the gradual development of human societies and cultures. The idea of progress from the simple to the complex, from the elementary to the sophisticated, was crucial to Pitt-Rivers's anthropological credo. And where supporting evidence was in short supply, he had no hesitation about engaging in conjectural cartography. Gradual improvement was no less central to his political thinking. Because he was convinced that institutional development was slow and steady, he recoiled at the thought of radical social change. It was by incremental evolution that the state, the family, and language had developed. And he hoped that in a period of chronic political unrest in Britain during the final quarter of the nineteenth century, his museum would convey to the public a salutary caution against insurrection. Nature exhibited no jumps—and neither should society. The Pitt-Rivers museum was thus a tract for the times on the benefits of moderation and the value of education. Gradual development was the order of the day in natural history, human institutions, and technological arts; and, he insisted, "this knowledge can be taught by museums, provided they are arranged in such a manner that those who run may read."

One of the most conspicuous expressions of how its exhibits were structured according to the museum's internal geography was the remarkable "sociological laboratory" spearheaded by Patrick Geddes in 1892 at his Outlook Tower in Edinburgh (fig. 8). This construction was intended to be a novel type of museum in which the study of civics and regional survey was encouraged through a synthetic integration of landscape, history, and sociology. Essential to this "temple of geography," as one of his peers called it, was the orchestration of its internal spatiality. The various stories were organized in a hierarchy. On top was the "prospect," housing a camera obscura through which the city of Edinburgh could be observed. Below this level was the Edinburgh Room, accommodating a scale model of the city with accompanying plans, maps, and engravings of its architectural heritage. The bottom two stories dealt with Europe and the world, respectively.

This interior geography was designed to lead the visitor from the local via the regional to the global. It was a dramatic spatial articulation of Geddes's entire philosophy of knowledge. Committed to social reform and global humanitarianism, Geddes persistently sought to equip people to engage in political transformation by fostering regional awareness. The Outlook Tower thus exuded its architect's sense of global holism, his commitment to educational innovation, his conviction that direct experience of the world should replace bibliolatry, and his assurance that regional particularity was the outcome of global evolutionary forces. It also imposed a hierarchical taxonomy on its exhibits and conveyed the impression that the world was composed of a nested set of interregional relations. The Outlook Tower,

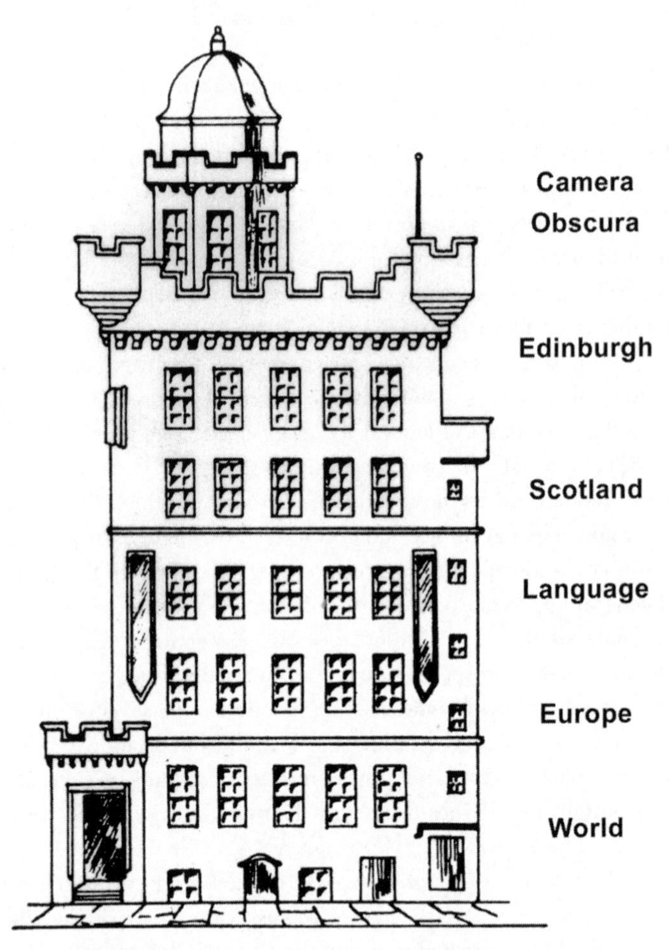

8. A diagrammatic elevation of the Outlook Tower in Edinburgh's Castlehill. In Patrick Geddes's conception, this sociological laboratory was organized around a nesting set of discrete but connected geographical spaces, each designed to lead visitors from the local to the regional and beyond to the national and the global. A device in spatial education, the Tower's internal geography was intended to mirror the world's geography at various scales.

which Geddes envisioned as an *encyclopaedia graphica,* thus constructed the world he hoped to reform.

It would be mistaken, though, to think that museums were just passive reflectors of their curators' preferences. Museums were not simply cartographic texts. They were often sites of struggle between curators, academics, sponsors, and the general public, all of whom had different aspirations for the institution. Moreover, the very materiality of the museum's physical space could "bite back." Its transformative influence was clearly felt on late nineteenth- and early twentieth-century anthropology in Berlin, where the study of the subject was conducted outside the university sector in the Royal Museum of Ethnology. German anthropology had been dominated by a craving to accumulate cultural artifacts, and this appetite was gratified through global networks of commodity acquisition. Such cupidity was theoretically justified by the belief that single items were nothing more than mere curiosities and that any particular article acquired anthropological value only when placed in a complete series. Collectors engaged in unlimited hoarding of specimens on the conviction that anthropology was best pursued through a comprehensive overview of humanity's material culture. Such thinking resulted in uncontrolled stockpiling in the Berlin museum. But the confines of the museum meant that the collection soon descended into chaos, lacking any order. The very ethnological overview that the museum was intended to supply was subverted by the volume of objects amassed. In due course, with disquiet being expressed by the public and alternative conceptions of anthropology developing elsewhere, curators eventually adopted new methods of handling artifacts. Intellectual change and spatial constraint went hand in hand. Ironically, the very site that had cradled the developing discipline provided the structural foundation for discarding the approach to the subject that it was originally established to advance.

If a museum's internal geography could condition the cognitive shape of the science produced, its external iconography could speak to the society in which it found itself. Museum architecture is not simply a set of structural answers to practical problems. It is itself a symbolic writing of space. The very buildings where scientific inquiry was housed were often pronouncements in the language of stone, site, and plan about the place science should occupy in the wider culture. We might allude, for example, to the ways museum architecture echoed ecclesiastical forms. Alfred Waterhouse's Natural History Museum in South Kensington was often referred to as "nature's cathedral." Opened to the public in 1881, this Gothic Revival "temple of science" was the world's most remarkable structure of its kind (fig. 9). Such celebratory ascriptions were entirely in keeping with the efforts of certain elements in late Victorian society to wrest social authority from the clergy and deliver it into the hands of a new scientific elite, even if the design itself was reflective of its director Richard Owen's personal enthusiasm for natural theology. After all, the scientific fraternity that congregated around T. H. Huxley, who saw himself as a "bishop" of the "new ecclesiology," sang "hymns to creation," preached "lay sermons," joined the "church scientific," and was ordained to the "scientific priesthood." In such circumstances architectural symbolism could well become one more weapon in the arsenal of cultural conflict.

While its architecture was intervening in the cultural struggles of late Victorian society, the museum as an institution did much to promote what has been called an "object-based" approach to knowing in the decades around 1900, not least in the United States. In a period when Chicago's Field Museum, the American Museum of Natural History, Harvard University's Peabody, and a host of similar institutions came into existence, the idea that knowledge could reside in material objects as much as in texts gripped the imagination of American intellectuals. Appropriately, apologists for museology urged that what distinguished their efforts from those of their antebellum predecessors was precisely that in the new museum specimens were viewed as objects of scientific scrutiny, not simply as spectacle. By thereby domesticating the dazzling, the fundamental order of nature could be unveiled by rational inquiry. The Philadelphia paleontologist Edward Drinker Cope sought to capture the spirit of the moment: "[As] the middle ages were the period of cathedrals, so the

9. The Natural History Museum in Kensington, with its cathedral-style architecture, encouraged visitors to see it as a temple of science. One commentator observed that visitors to this "animal's Westminster Abbey," with its stained-glass windows and churchlike atmosphere, were known to respectfully remove their hats on entering the building.

present age is one of colossal museums, and of an extensive development of knowledge of the sensible creation." By giving such priority to objects and their value, the American museum fit remarkably well into a culture of acquisitiveness and, for a short time at any rate, was in the vanguard of the new century's cultural economy of science.

This triumph was short-lived. The modern research university soon acquired the cognitive authority that had resided in the museum. At the same time, the realization that the meaning of artifacts is unstable and shifts depending on how objects are arranged, tended to downgrade their scientific significance. For all that, the late nineteenth-century museum constituted a remarkable experiment in visual encyclopedism.

The museum, it is clear, has performed a variety of roles in the historical unfolding of scientific inquiry. Occupying a distinctive niche in the ecology of science, it constitutes a space where items have been accumulated and allotted their "proper place" on the stage of history. In this way museum culture played an important part in the history of "viewing." In the museum people learned how to look at the world, how to value the past, and how to visualize relations between specimens. Yet no matter how extraordinary the exhibit or how dramatic the diorama, the museum was not the world itself. The museum was no mirror of nature. To view *that* required moving outside the confined spaces of the collectors' cabinets and into the open spaces of the field—another site of scientific endeavor.

Field Operations

The idea that the world should be its own laboratory, and that the best way to study some part of nature is to go there and experience it firsthand, is anything but the obvious claim it appears to be. When the French comparative anatomist Georges Cuvier commented on the scientific travels of Alexander von Humboldt in the early nineteenth century, he sharply contrasted the styles of scientific travelers and "sedentary naturalists." Because the former quickly traversed territory and viewed many things in sequence, Cuvier insisted, they could "only give a few instants of time to each of them." The observations of the fieldworker were "broken and fleeting." By contrast, the bench-tied student of nature had the time to spread out samples, to collate and analyze them, and thereby to come to reliable conclusions. The laboratory naturalist occupied a kind of hyperspace: because creation in all its dazzling diversity passed across the workbench, it afforded the opportunity to rearrange the natural order and grasp it as a whole. By patient comparison and correlation, the armchair naturalist could easily triumph over the fragmentary and precarious claims of the fieldworker. For Cuvier the most wonderful voyages of discovery never weighed anchor and pushed out to sea: they never left the workshop. Only in the study could one rove the cosmos.

Whatever the merits of Cuvier's partisan analysis, his interventions call attention to the markedly differing cognitive styles that characterized open-space and closed-space naturalists. For the former, as one commentator puts it, "mastery over and comprehension of nature" derives from "*passage over* terrain"; for the latter, "the steady and immobile *gaze*" is accorded cognitive privilege. Cuvier's conviction was deeply ironic. The very thing that secured the reliability of the sedentary naturalist was what advocates of field science strenuously repudiated—*absence* from wild nature. To fieldworkers it was *presence,* not absence, *closeness,* not distance, that underwrote their claims to authenticity. Their immediate experience of moving through space, often heroically, with all the bodily exertion and rigors that entailed, provided warrant (as we shall see in chapter 4) for the scientific stories they had to tell. Dissecting specimens and displaying exhibits were all well and good, but it was only in the field that nature could be encountered in the raw. The workshop bench could deliver only a virtual world—valuable enough, but no substitute for the real thing.

Nor was Cuvier's dispute with Humboldt a unique episode. To the mid-nineteenth century Edinburgh student of Alpine glaciers, James David Forbes, it was only "protracted residence among the Icy Solitudes" that warranted genuine scientific knowledge of glacial matters. It was only presence in the ice fields that could replace rumor with reason. The Cambridge mathematical theorist William Hopkins, however, didn't see things the same way at all. To him the nature of glacial motion could be deduced from the laws of physics and their operation in laboratory-based experiments on force, solids, and fluids.

What was going on here, fundamentally, was a dispute about appropriate modes of scientific knowing. And that debate was not without its fair share of name-calling. Field men like Forbes, and indeed the Irish physicist John Tyndall (whose views on glacial motion differed from Forbes's but who shared his manly enthusiasm for heroic rigor in remote places), did not hesitate to dismiss their opponents as mere armchair theorists. The rhetoric of adventure dominated the culture of field science: adventurousness conveyed its own authority. Laboratory opponents, by contrast, felt that high adventure in an uncontrolled wilderness delivered nothing like the precision good science demanded. Fun was one thing, physics something else.

As these two debates reveal, enthusiasts for field science regularly appealed to location as a key component in justifying their knowledge claims. To them *where* science was practiced constituted an important strand in their arguments about why they should be believed. Credibility was, to a considerable degree, a matter of locality. And yet just exactly what "the field" means has never been clear-cut: it is shot through with ambiguity and inconstancy. As an open space it is less easily defined, bounded, and policed than its intramural counterparts like the laboratory or the museum. For this very reason the field is inhabited differently from these other scientific spaces. For a start, the investigator here is likely to be the visitor rather than the resident—precisely the converse of the laboratory world. The settled inhabitants of the field site are not the scientific experts engaged in research. And there are likely to be other transient sojourners such as tourists, campers, foragers, artists, and hunters, to name but a few. The variegated nature of the field's dynamic human geography makes for an unstable network of social relations. The field thus discloses precisely the kind of sociology that the laboratory seeks to escape, with its formal and informal disciplines geared to maintaining stability.

In these and other ways, the field is a space where the structures of social life are at once reproduced and destablized. The ambiguities of presence and absence are also significant here. Take, for instance, the involvement of amateurs in field sciences. Although they are fundamental to everything from archaeological digs to botanical surveys, their presence has been regarded as cognitively compromising by those promoting the supposed rigor of laboratory standards. And indeed, while the boundary between the professional and the amateur is much less clear-cut in the field than elsewhere, it is true that "amateur knowledge" often has passed as genuine science only when warranted by the credentialed professional.

A similarly equivocal position has been occupied by women in the field. On the one hand, the field has often been promoted as a manly site of intrepid heroics, with its narratives cast in an epic form that celebrates the virtues of stoicism, resilience, pragmatism, and inventiveness. Indeed, the impression has sometimes been given that these practices make their own contribution to justifying claims to knowledge. Often, too, fieldwork has been venerated as a rite of passage that the fledgling scientist must struggle through, both metaphorically and literally, to achieve mastery over nature. These values have not always been attractive to women. And since the field was foundational to such Victorian sciences as geology and physical geography, it may well have colluded in their exclusion from these groves of the academy. On the other hand, the foreign field has sometimes afforded women the opportunity to escape from the rigid regimens of the homeland. Their personal experiences far away, moreover, occasioned domestic equivocation. When Mary Kingsley (whose *Travels in West Africa* appeared in 1897) returned to England from her second journey to the "dark continent" in 1895, she was welcomed by the press as a marvel, a novelty, a wonder—"a lonely English lady" who had "manfully" borne the trials and tribulations of the foreign field. Plainly, her conduct abroad transgressed the virtues of her comportment at home. She might be a heroine, but she was also an anomaly.

Women also energetically participated in the domestic field club movement in Victorian England, which did much to foster amateur science at the time. At least in part a manifestation of a romantic sensibility toward the natural order, groups like the Berwickshire Naturalists' Club, formed in Scotland in 1831, opened membership to both men and women. Such societies were as much experiments in innovative social relations as places where the cult of the naturalist was actively nurtured. By the same token, middle-class women who did infringe the conventions of gender relations by going on such field outings did so—emphatically—as amateur "botanophiles," not as professional botanists. Indeed the amateur/professional polarity could itself operate as a means of excluding women from serious scientific visibility and underscore the presumption that, for women, natural history was nothing more than a genteel hobby.

Ambivalence also attended the remarkably successful field ventures of Alfred Russel Wallace (the codiscoverer of evolution by natural selection) in the Malay Archipelago during the nineteenth century. However transforming a personal experience it was, and however much his ventures were presented in the language of the rugged explorer, Wallace's time in the East depended on an existing colonial network of merchants, government officials, medical practitioners, and clergy into which he easily fit. To be sure, in its transplantation to the colonies, the structures of British society had undergone change. Relations between the middle and upper classes were less fixed than at home, for example, and Wallace used this social fluidity to good effect. Nevertheless, like Humboldt in South America, Wallace was able to rely on the entangled systems of socioeconomic interchange that European colonists had woven. Intrepid his excursions may have been; isolated they certainly were not. For in order to overcome the distrust of the local peoples he moved among, he had to depend on the friendship, loyalty, and trustworthiness of a variety of associates to procure the information he sought. Here the idiom of the heroic individual is all wrong. Wallace's field science was an inescapably social affair. And the knowledge he acquired was the compound product of personal observation, trusted testimony, and colonial infrastructure.

At once restrictive and liberating, by virtue of its social flexibility fieldwork offered greater space for renegotiating personal and vocational identity. The field allowed scientists scope to engage in resourcing the self. Whether breaking down gender roles, encouraging the transgression of social conventions, blurring the line between amateur and professional, or furthering the mythology of hardy heroism, the field regularly exhibited a borderland sociology and a frontier mentality. While these arose in some measure from the *human* geography of the field's occupants, its *physical* geography has also played its part. The field is an inherently unstable scientific site, and for that very reason practical rationality and functional imagination are at a premium there. Local conditions pose local problems needing local solutions. In such circumstances science is an inescapably local practice. Here the good scientist is the skilled hand, the resourceful artisan. Not that such aptitudes are irrelevant in the laboratory. But in the field, replication is not so easily effected, the environment is less readily controlled, and impromptu ingenuity is in correspondingly greater demand. And yet however innovative in situ practices may be, the crafts deployed in the field are nonetheless typically acquired at home. Encounters with the unexpected are routinely construed in customary ways, for field scientists—it has tellingly been said—"travel with their domestic habits of mind and behavior." And this is not only the case for acquiring field knowledge; it is no less important to communicating findings. The singular experiences of the field can be expressed only by using a common lexicon and drawing on shared cultural resources. To that extent the homeland is always present with the scientific traveler (fig. 10).

That science is a cultural *practice,* then, is exemplified with particular clarity in the field. For here hands-on experience, routine improvisation, and performative rationality are highly valued. The caliber of the science produced is a direct reflection of the quality of practical reasoning and proficiency at manipulation. This reminds us that rationality is not independent of the customs and practices that constitute a tradition of inquiry. To the contrary, it is embedded in them. Theory and practice need to be thought of in reciprocal rather than oppositional terms. It is in practices as much as in theories that traditions of inquiry articulate themselves, and they do so in activities that are not reducible to formal rules of inference. There is no scientific rationality that is independent of a tradition's procedures, customs, and performances—that is, of the practical conditions of knowledge making. For this very reason what we might call "ap-

10. This 1849 sketch by William Taylor shows the botanist Joseph Dalton Hooker in the Himalayas receiving rhododendrons from local people as colonial tribute. The imperial mind-set that Hooker brought with him from home continued to find expression in his foreign fieldwork. Indeed, it animated his passion to keep Kew Gardens at the center of a colonial scientific network that stretched around the globe.

prenticeship" is essential in the field sciences, where one can learn how to deal with the exigencies of the contingent only by working under the authority of an accomplished practitioner.

In some measure at least, the centrality of practice and the premium put on the craft competencies of the fieldworker arise from the open space and deeply uncontrollable character of the field. But it would be a mistake to think that the field is simply a site that just "is there" and can be taken for granted. Rather, it is constituted "as the field" by the activities of scientific investigators. Because of the power the academy has to define the field and thereby, in many cases, to justify its own "field of inquiry," the field site is always politically negotiated. In some academic disciplines, notably anthropology, fieldwork has been a kind of fetish that has normalized the domain's practices, empowered certain styles of knowledge while impeding others, and sanctioned certain objects of study. In anthropology it was Bronislaw Malinowski who installed fieldwork as central to the institutionalization of the discipline. Thereby he effected a move away from the worldview of Victorian gentlemen-scholars who considered going into the field rather beneath their dignity. Courtesy of his organizational skills, the field methods Malinowski had deployed in the Trobriand Islands rapidly became the legitimating insignia of the profession. By valorizing the field, the new professionals were able to subvert the authority of the old gentlemen-naturalists. From then on, fieldwork became "the central ritual of the tribe."

In important ways, then, the field is constituted by academic projects and narratives. Its existence as a scientific site depends on the stories scientists tell about it. While this is the case with all field sites, it is in the social sciences that we may most clearly catch a glimpse of how this is so, for here the relationships between the researcher and the researched dramatically surface. The investigator has the power to group individuals into some abstract collective and then label them as slum dwellers, domestic servants, middle-class fundamentalists, migrant workers, or some such. This is because the social scientist delineates the boundaries and defines just who is in and who is out of the subject circle. The politics of fieldwork thereby surface as part and parcel of the politics of representation. The knowledge claims arising from fieldwork in the social sciences are thus intensely local. And they are local in two senses. First, the information collected is about circumstances in some particular locality; second, the entities that social theories seek to explain are constituted by the analytical categories the field investigator imposes on local data.

In other ways too, fieldwork gives voice to the political commitments of the researcher. Late nineteenth-century urban fieldwork, for example, often had the aim of rendering service to the dwellers of disadvantaged areas. More recently inner city "expeditions" have been mounted by radical social scientists to advance what might be called an emancipatory geography that is designed to empower the marginalized and enable them to escape from the grip of oppression and the spiral of poverty. Fieldwork in such scenarios is a prelude to political liberation and a chapter in its history. Frequently in these cases relationships are further complicated because the conventional distinction between "home" and "field" does not apply as "insiders" and "outsiders" are elided. In this sense the field is a space that is at once familiar *and* foreign.

The field, then, turns out to be anything but the obvious scientific site it might initially seem to be. Characterized by ambiguity and constituted by academic projects, fieldwork has nonetheless been installed as an operational answer to questions about appropriate ways of knowing for certain traditions of scientific inquiry. Absence from home and presence in the field, as the necessary precondition of bona fide knowledge, was the outcome of historical negotiations that gave the field sciences their distinctive place in the scientific division of labor. Here cognitive warrant was built on the foundations of spatial practices, for fieldwork literally grounded the claims of the scientist.

Gardens of Display

Between the archive and the field, the world of the museum and the world of nature, stands the garden. A site of botanical and zoological inquiry, the garden has a complex spatial history in which different purposes and practices have intermingled. It is a multilayered space whose meaning has undergone manifold transformations, each trailing clouds of earlier associations. Enclosed yet expansive, open yet delimited, natural yet managed, the garden occupies a place between the great outdoors and the cloistered cabinet. It was always so. Wasn't God the first gardener when he planted the Garden of Eden? It was a spiritual site in which its human inhabitants walked with their Creator. But once sin entered their lives they were expelled from its pleasures and perfections. Since then, in the Christian tradition at least, every gardener's battle against the encroachments of the wilderness has been an attempt to reflect, if not retrieve, primordial paradise (fig. 11). From earliest times, the garden has been seen as a place of retreat and renewal, an outdoor temple of contemplation and meditation in which spiritual well-being could be maintained. Further, the garden's very existence has depended on its capacity to represent order over against chaos, cultivation in opposition to wildness, art as opposed to nature. The boundary of the garden thus marked out a line between the rational and the irrational. As a space of display, the garden was meant to present the orderliness of creation by recovering Eden's pristine harmony. Not surprisingly, the garden long remained a fertile repository of ecclesiastical metaphor and spiritual allegory. It was a "type" of heaven, with trees an emblem of Christ, branches a symbol of the saints.

With scientific pursuits these meanings began to be reshaped. Whereas early gardeners yearned for the recovery of ancient wisdom, often in the hope of retrieving the lost powers of Adam, scientific travelers lusted for new knowledge. In the wake of the European voyages of reconnaissance, the conception of the garden as a hallowed refuge from the world began to be supplemented by a vision of the garden as a living encyclopedia. As plants arrived from across the globe, they were identified, named, and allotted their proper places in the garden's spatial taxonomy. The early botanical garden was both a re-creation of paradise and a key moment in the genesis of modern science. Even as the encounter with the New World challenged the

11. Seventeenth-century gardeners sought to re-create the kind of earthly paradise depicted in this 1629 representation of the Garden of Eden by John Parkinson. Gardening could thus be at the same time a scientific, medical, and theological pursuit.

classification schemes of the ancients, it no less inspired the hope that, for the first time since the fall from grace, the plenitude of Eden could be restored. The seventeenth-century author Abraham Cowley, for example, insisted that America had brought back into view lost elements of the creation and that Eden could be re-created by reassembling in one location the scattered pieces of the globe's plant jigsaw puzzle. The first modern botanical gardens, established in Padua and Pisa in the early 1540s and, for the English-speaking world, at Oxford in 1621, thus served the interests of both theology and science.[1]

This was clearly so in the collection of the mid-seventeenth-century gardener John Tradescant, which housed a rich variety of items. Widely known as "the Ark," it reinforced connections between the biblical Noah and natural history. This association was intentionally optimistic. It was in Noah's ark that God had restored human dominion over the creation, and it was thought that its replication could recover optimal conditions for acquiring reliable knowledge. The ark, after all, was God's museum laid out according to divine specifications. Gardeners like Tradescant were latter-day Noahs engaging in a task of spiritual and scientific retrieval and reversing the global effects of negligence and depravity. The ark, like such other biblical sites as the Garden of Eden and Solomon's temple, provided the seventeenth century with images of ideal knowledge spaces. The temple, for example, provided inspiration for the restoration of a godly society in which cooperation and diligence would yield true understanding. Here the cognitive effects of Adam's fall could begin to be reversed. Clearly, the site of knowledge acquisition was crucial to establishing the integrity of the knowledge procured.

Understandably, the garden's internal geography began to be rethought in consequence of its rapidly growing range of specimens. The layout was meant to map onto the globe in some discernible way.

1. Other early botanical gardens include Zürich, established in 1561, Lyons in 1564, Rome in 1566, Bologna in 1567, Leipzig in 1579, Leyden in 1587, Montpellier in 1592, Heidelberg in 1593, Giessen in 1605, and the Jardin des Plantes in Paris in 1635.

The four continents were each allocated their literal "quarters" in the garden. John Hill, for example, specified—in his *Idea of a Botanical Garden in England,* which appeared in 1758—that the sections should be "appropriated to the four great regions of the earth, and defined for the reception separately of European, African, American, and Asiatic plants." In eighteenth-century France, the landscaping of botanical gardens was carried out to provide what one historian has recently called "simulacra of different climatic and topographical conditions." By geographical planting, as it was called, the garden was intended to display the elegance and symmetry of global botany. Not that it always did so with identical design arrangements. Some used circles, some squares, some circles enclosed in squares, or a dozen other variations (fig. 12). Either way, the prodigality of the natural order was systematically tamed by symmetrical reconfiguration, its blithe randomness brought under the reign of enlightened rationality. The garden also reduced the global macrocosm to a microcosm, to what Francis Bacon called "a model of the universal nature made private."

What remained dominant was a fascination with geometric precision and proportional symmetry. And this, as often as not, reflected the belief that God had laid out the first garden in an orderly fashion, so unlike the chaos and confusion of the postlapsarian world. This was not invariably the case, of course. Seventeenth-century French formal gardens used ever more sophisticated geometrical arrangements to express the economic success of mercantile capitalism. As an item of conspicuous consumption or ostentatious exhibition, these gardens declared their owners' social station in a period of burgeoning elite consumerism. In so doing they moved nature from the domain of divine creation to secular property. As well as being sites for accumulating botanical specimens, formal gardens were maps of both social status and buying power.

Gardens could recover paradise. They could give decorative expression to economic might. And they could also be instrumental in reversing the ravages of the biblical fall from grace by releasing the medicinal powers embodied in its specimens. Spiritually, aestheti-

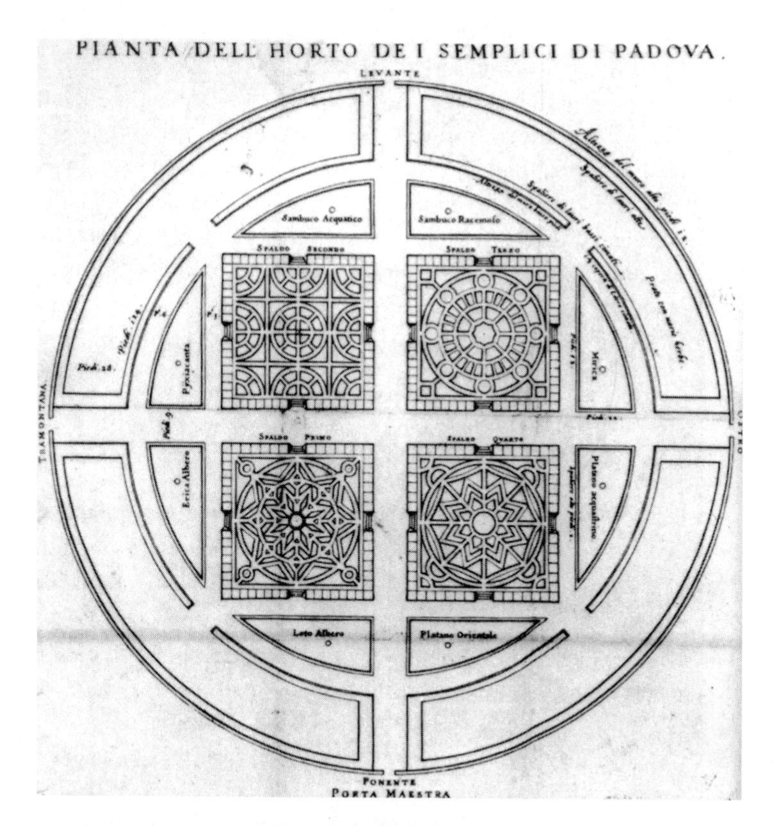

12. Plan of the botanical garden at Padua (1591) by Girolamo Porro, who asserted that at Padua the whole world was being gathered into a single chamber. The aim was to assemble specimens from the four quarters of the world and allocate them to their proper places in the garden's schema.

cally, and now medically, the garden was an exercise in restoration. John Evelyn, in the mid-seventeenth century, asserted that gardening was an empty occupation unless graced with some tinge of medicine. Not surprisingly, the first "physic" gardens, as botanical gardens were often called, flourished in the medical faculties of universities, at least in part to shield apothecaries from unscrupulous traders in drugs and roots. Associated teaching positions in what was referred to as the "simples" were established to identify the curative properties of plants and to recover long-lost botanical-medical lore. The craft of the pharmacological botanist frequently involved reading the "signatures" of the vegetable world so as to specify which part of the body each plant was designed to treat. Walnuts, for instance, were understood as having the sign of the head, with an outer husk that looked like a skull, and therefore embodied substances suitable for treating head wounds. Because plants possessed virtues that could be released, the search was on to extend herbal knowledge to the newly encountered plants arriving from across the oceans. In this way the study of medicinal botany conferred on its practitioners power over nature and people alike. And gathering global plant riches into one space—the garden—was the best way of acquiring this power.

Botanical gardens, then, were multifarious spaces. They hankered after the Garden of Eden; they sought to reproduce global biogeography; they exhibited social standing; they wielded biomedical power. Whether as stages for the display of courtly ornamentation or as symbols of royal glory, as temples of divine contemplation or as theaters of useful natural philosophy, botanical gardens touched on the deepest needs of the state. Given these preoccupations, it is not surprising that it became increasingly fashionable to resort to political metaphors to describe the plant world. Plants were thought of as nations, each with its own provinces and member species. Such Enlightenment naturalists as Johann Reinhold Forster, Eberhardt Zimmermann, and Alexander von Humboldt treated plant associations as if they were political entities, and their methods of study were precisely the same as those that statesmen used in their political arithmetic. They were concerned with the social economy of the vegetable world.

Such political analogies flourished with particular vigor during the eighteenth- and nineteenth-century age of empire, when metropolitan gardens became the hub of botanical imperialism. Kew Gardens, for example, whose origins can be traced back to the 1750s, burgeoned under the vegetative booty brought back by men like the eighteenth-century botanist and scientific statesman Joseph Banks and his collectors, who engaged in worldwide horticultural plunder. Interest in the profit to be derived from economic botany, of course, was not all one-way. As in Amsterdam, crops were cultivated at Kew for export to the colonies. Thereby the gardens furthered the commercial vitality of the nation. Indeed, from the mid-1780s it became the center of a worldwide network of plant acquisition and exchange, a nodal point in what has been called the "Banksian" empire. Thousands of seeds, plants, and dried specimens, some covertly pillaged for commercial gain, others as mere instances of exotic curiosity, found their way to the ecumenical data bank at Kew. From Southeast Asia and the Pacific to the West Indies and Central America, an intricate system of plant trade came into being with the intention of harvesting the economic riches of Banks's botanical empire. Hemp seeds, tea plants, mulberry, natural lacquers, tung oil, fiber plants, citrus, avocado, and myriad other items were sought as diet and drug, dye and decoration. As Banks exploited his contacts with diplomats and navy officers, missionaries and tradesmen to garner the world's plant riches, Kew took delivery of a nutmeg tree and mangosteens from Java, plants by the boxload from Canton, Tahiti, Tasmania, and New Guinea, and packages of seeds from India. Regulating the botanical traffic that flowed back and forth between metropolitan core and colonial periphery, Kew enriched the fiscal and scientific capital of the empire. And to sustain the industry in horticultural cargo, satellites of Kew (often directed by Kew-trained curators) developed in such places as Jamaica, St. Vincent, St. Helena, Calcutta, and Sydney.

If botanical gardens were agents of empire, they were no less sites of experiment and enlightenment. Whether tropical plant species could acclimatize to temperate zones, and vice versa, was a scientific question as important to imperial success as to intellectual progress. Precisely because Kew Gardens was one of the great exchange houses of the empire, it became a testing ground for trials in botanical acclimatization, a project in remaking nature to suit the new industrial order. At the instigation of Banks and his collaborators, varieties of hemp and flax, trees and vines, fruits and vegetables all crisscrossed the globe in hopes of adapting them to new climatic regimes. The botanical garden was often their first port of call. Plant collections were a prime location in the pursuit of useful knowledge, and under Banks's influence Kew became the archetype of Enlightenment botany. As he himself noted with pride, "Our King at Kew and the Emperor of China at Jehol solace themselves under the shade of many of the same trees and admire the elegance of many of the same flowers in their respective gardens." Such pleasures attested to the runaway triumphs of botanical acclimatization.

But it was not just plants that were the subject of such inquiry. Precisely the same questions arose over animal trafficking. And discovering how animals adjusted to new climatic conditions (if they did) often became the opening gambit in campaigns for the creation of modern zoological gardens. Whether animals belonged in botanical gardens was a long-standing and unresolved dilemma. Thomas Aquinas, for example, suspected that after Adam had named them, the animals were excluded from the Garden of Eden, whereas Basil and Augustine insisted that their presence gave Adam and Eve much pleasure. So while some sought to keep the botanical garden free from animal intrusion, others thought it should incorporate all aspects of creation, and they therefore supplemented geographical planting with animal representations of the continents—the zebra for Africa, the llama for the Americas, and so on. In consequence, many royal botanical gardens such as those at Kew and Versailles housed small menageries of exotic creatures.

Insofar as zoological gardens were bound up with animal domestication and acclimatization, they were invariably implicated in colonial projects (fig. 13). Three nineteenth century zoos—in Britain, France, and Australia—nicely illustrate this association. When Stamford Raffles, founder of the Zoological Society of London, returned in 1824 from his imperial adventures in the East, he was irked to find that Britain was lagging behind other European nations in matters of zoological display. Despite its glorious global empire, Britain's facilities for exhibiting exotic animals amounted to little more than fairground sideshows and frivolous entertainments—mere spectacles for titillating the vulgar—not to be compared with the "mag-

13. Isidore Geoffroy Saint-Hilaire is shown here conducting the Empress Eugénie and the prince imperial around the tropical house at the Jardin Zoologique d'Acclimatation. Such associations were symbolic of the garden's imperial ethos.

nificent institutions" of its Continental neighbors. To relieve this cultural embarrassment in a manner befitting the grandeur of an imperial power, Regent's Park Zoo opened its gates in 1828. When addressing its landowning constituency, the zoo rationalized its existence by stressing its concern to domesticate exotic species and acclimatize them for English parks—and menus. After all, the anatomist Richard Owen of the British Museum and the London naturalist Frank Buckland, both enthusiastic advocates of acclimatization, later organized "exploratory and adventurous" dinners to support their obsession with domestication. Such delicacies as kangaroo steamer, Honduras turkey, Syrian pig, and tripang soup made from Japanese dried sea cucumber all featured. These dinners were experiments in gustatory geography. When advertising its wares to the scientific community, by contrast, the zoo presented itself as a reservoir of taxonomic data without reference to table fare or ornamentation. The zoo thus existed in the shared space between applied natural history and Linnaean science. Either way, the vast array of specimens displayed in the zoological gardens served to draw attention back to Britain's ecological imperialism. The zoo, we might say, was a rhetorical site of empire, its animals intended to symbolize Britain's biogeographical dominance of the world. The globe, it seemed, existed to serve Britain—gastronomically as well as scientifically.

Much the same was true of the Paris collections. Three generations of the zoologically inclined Geoffroy Sainte-Hilaire family were vital here. Étienne founded the menagerie at the Paris Museum of Natural History in the 1790s, his son Isidore promoted the Jardin Zoologique d'Acclimatation, and the grandson Albert succeeded to the directorship of the Jardin for nearly thirty years. In one way or another, these various collections reflected the country's colonial, diplomatic, and commercial activities, especially in North Africa; and as the century wore on, the acclimatization garden came to enjoy the imperial patronage dispensed by Napoleon III. As in Britain, here too there were tensions between pure and applied zoology. The pendulum swung from utility to science at different times, and between the needs of the naturalists and the amusement of the general public. Either way, breeding and dealing in exotic animals were seen as contributing to agriculture and industry, scientific advance and commercial success alike.

The French scientific community had had a long-standing interest in acclimatization, not least because it bore directly on matters of adaptation, inheritance, and evolutionary change. In fact by the mid-nineteenth century the Jardin, which originated as a royal physic garden, was in large measure the public laboratory of the Société Zoologique d'Acclimatation, which had come into being in 1854. Successful long-term adaptation of species to new environmental niches would do much to confirm the doctrine of the inheritance of acquired characteristics and thus the biological transformism rooted in the earlier ideas of Georges-Louis de Buffon, who had been intendant of the Jardin for nearly fifty years, and Jean-Baptiste de Lamarck, zoology professor at the Muséum d'Histoire Naturelle, its

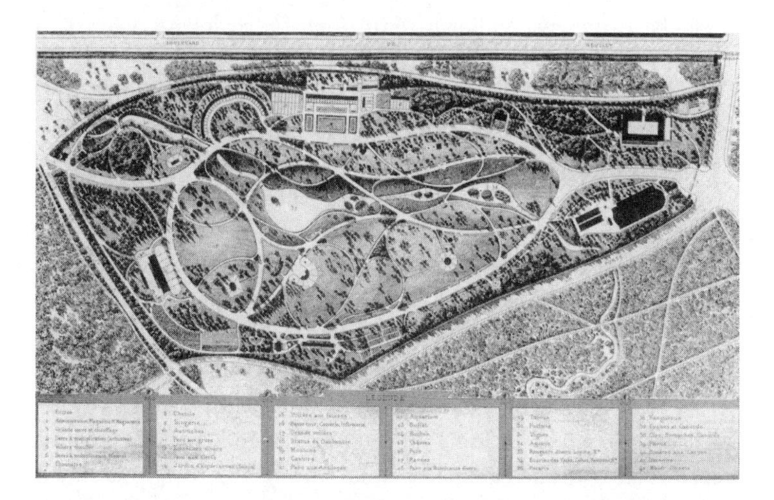

14. Plan of the Jardin Zoologique d'Acclimatation, which opened in Paris in 1860. The Jardin presented colonial France's faunal resources and included such exotic specimens as Moroccan wild sheep, Angora goats, and Tibetan yaks. Only exhibits of "public utility" were to be displayed, and the concern was to ensure that they could acclimatize to a new environment. At the Jardin, scientific inquiry was intimately bound up with matters of imperial practicality.

successor institution. Yaks from Tibet, wild sheep from Algeria, Angora goats, Egyptian ibis, and llamas from Chile, when gathered into zoo space and appropriately displayed, could advance French science, proclaim the nation's colonial splendor, and help visitors conjure up an imagined round-the-world safari (fig. 14).

Acclimatization also had a central role in the genesis of the Melbourne Zoo, not least through Edward Wilson, an English-born editor of the Melbourne newspaper the *Argus*. Wilson was fully aware of the acclimatization projects of the Parisians, and thoroughly impressed by their ideals, he began a public campaign in the late 1850s for introducing new plants and animals into British colonies. More particularly, he passionately believed that Australians had a right to the ornithological pleasures and hunting thrills of old England. It was utterly shameful, he felt, that despite its geopolitical domination of the globe, Britain was failing to redistribute the ecological riches of its empire to its far-flung colonies. And so he embarked on an untiring campaign to establish colonial acclimatization societies. Thanks to his efforts and those of Thomas Embling, a medical doctor and political activist, an experimental farm and zoological society emerged in 1857. Though it was short-lived, it eventually resurfaced as the Royal Melbourne Zoological Gardens.

It would be mistaken to imagine that modern zoos owe their existence exclusively to the post-Enlightenment fascination with acclimatization. For a start, menageries were in existence as long ago as 2500 BC in Egypt; Ptolemy, in the third century BC, founded a zoo in Alexandria; in ancient Rome *vivaria*—animal holding zones—were available for public scrutiny; and the Aztec emperor Montezuma kept a great aviary and animal enclosure. Moreover, royal households routinely collected and exhibited exotic creatures as a mark of prestige and power, and during the sixteenth century numerous menageries surfaced in the rising urban centers of Europe and North Africa—Prague, Karlsburg, Constantinople, and Cairo—at least in part as a mark of civic pride. The menagerie of Versailles, it has been said, was "first and foremost a political testament to the power and majesty of the king," Louis XIV.

Nevertheless, the efflorescence of zoological gardens in the nineteenth century owed much to the intellectual and commercial potential of acclimatization-related matters. And there are grounds for suspecting that these preoccupations were not isolated from related anthropological questions about the effect an alien climate would have on human colonial populations. That such obsessions were never far from the minds of zoo magnates is clear from the incorporation of ethnographic exhibits into leading nineteenth-century zoos. Carl Hagenbeck, famous for his development of the zoo "panorama," in which animals came out from behind bars and inhabited open spaces, introduced what he called "anthropological-zoological" exhibits into his Hamburg Tierpark in 1874. That year he had Lapps acting out daily life with reindeers before enthusiastic audiences. Over the following half century he orchestrated some seventy ethnographic performances, Oglala Sioux performing ritual dances in the shadow of a constructed mountain being among the most popular. Similarly, Albert Geoffroy Saint-Hilaire enlisted caravans of Nubians, Canadian Inuit, and troops of Argentinian gauchos in the hope of maintaining public interest in his Jardin. And in 1906 an African Pygmy named Ota Benga was put on display in the Monkey House of the New York Zoological Park.

Such scientific "staging" of human subjects had profound ramifications, not least when such enterprises were replicated in colonized societies. In nineteenth-century India, for instance, the Asiatic Society proposed to the government that ethnological exhibits be appended to the General Industrial Exhibition of 1869–70. The proposal that aboriginal peoples should be displayed was made partly on the grounds of their anthropological peculiarities, partly because they would make good laborers around the exhibition grounds. All this had an unforeseen double effect. By showing such "specimens" as "types of man," it queried the normalcy of European conceptions of the human race. At the same time it distanced the intellectual elites of both Europe *and* India from those irrational bipedal members of the species.

The zoo, then, sometimes presenting itself in the metaphorical shape of the laboratory, took on the dimensions of theater. In so doing it also renegotiated the boundary between the animal and the human, the spectacle and the spectator, the viewer and the viewed, the rational and the wild—a boundary line that followed the contours of what was considered strange, exotic, peculiar, outré, other. The zoo thereby became a space reinforcing the profound sense of difference between exhibits close to nature (both animal and human) and visitors above nature.

Seen in this light, the zoo emerges as both a scientific and a theatrical space. The African Plains exhibit at the New York Zoological Society in the 1940s, for example, was little short of a simulated safari. Here a tribal village—surrounded by everything from warthogs to zebras—was re-created to instill a sense of adventure. Soon other wildlife parks were defending their existence on the scientific claim that exotic species had to be studied in their natural habitats. To achieve that, big game had to be enclosed by invisible, but no less real, steel fencing. In this guise the zoological garden fused the functions of field station and open-air stadium. Besides, the zoo was also a space of domination. By imposing order on the animal kingdom, organizing its exhibits along a rigid linear pathway, and caging dangerous large carnivores a tantalizing arm's length away, the nineteenth-century zoo testified to human triumph over the wild. Zoos, it has been said, "reenacted and celebrated the imposition of human structure on the threatening chaos of nature." The keeping and showing of wild animals was simultaneously emblematic of human power over the natural order, of metropolitan control over peripheral territory, and of imperial dominion over colonial empires.

Botanical and zoological gardens occupy a distinctive niche in the ecology of scientific practice. Spaces both of experimentation and exhibition, and open to the gaze of public visitors, they nonetheless accommodated inquiries very different from those carried out in the laboratory and the museum. Whether as agents of empire, sanctuaries of contemplation, or theaters of art, whether as symbols of power, reservoirs of medication, or maps of knowledge, gardens embody their own distinctive spatial formations of scientific knowledge.

Spaces of Diagnosis

Like the museum, the garden, and the zoo, the hospital stands both within the worlds of science and public culture and between them. Here, however, the concern is with diagnosis rather than display, with delivering care rather than accumulating objects. Still, it is only in the past century or so that the hospital has come to enjoy whatever positive images these pursuits convey. Before the twentieth century, most health care was dispensed in the home, and the hospital was disparaged as a site of destitution and danger. It was a risky space occupied by the feckless and friendless who had no option but to place themselves under the care of strangers. Indeed, in France and Italy these places had more general correctional purposes. Paupers, petty criminals, and prostitutes, among many others, shared space with the ill and infirm. It was not until the end of the eighteenth century that efforts were being made to disentangle the *hôtel Dieu* for the sick from the *hôpitaux généraux* for social outcasts of various stripes. The history of the modern hospital can be traced back to the monastic infirmary, the almshouse for the hopeless, army barracks adapted to tend the wounded in wartime, plague houses, and various other institutions that from time to time had to care for the sick. Modern hospitals, it is clear, had to carve their own specialist space out of existing establishments that happily mingled spiritual discipline, forced labor, psychotic restraint, cold charity, and treatment of the ill.

At the same time they had to work hard to shed the image of contamination and corruption in order to dispel the fears of those who condemned hospitals as pestilential swamps exhaling illness. After all, in late eighteenth-century Paris hospitals had beds occupied by as many as half a dozen patients, with little effort to keep separate those with contagious diseases, women in advanced stages of labor, and the dying. To overcome negative images of this sort, physicians themselves made such radical proposals as demolishing hospitals every half century or so to interrupt the cycle of disease transmission because, as one observer noted in the mid-1870s, hospitals did more harm than good. Sir James Simpson, who introduced chloroform anesthesia into Victorian medicine, even coined the term "hospitalism" to describe these morbid conditions. He once remarked that in hospital surgery patients were "exposed to more chances of death than was the English solider on the field of Waterloo." Given that Victorian surgeons operated on sawdust-covered tables in old, blood-encrusted garments and generally washed their pus-smeared hands only *after* surgery, it is small wonder that the rich elected to stay at home when sick.

As the meaning of hospital space moved with social judgment, so its changing architecture mirrored shifts in disease theory. In the nineteenth century, the miasmic theory dominated, postulating that disease was a consequence of noxious emanations. The matter of air flow through wards was understandably paramount. Florence Nightingale, for example, was preoccupied with ventilation because she was convinced that infectious diseases moving through the air could pollute a hospital's entire atmosphere. And her *Notes on Hospitals* of 1859 cited grim statistics to support her case. Hospital designers became obsessed with countering miasmas and giving priority to open spaces and patient isolation. The so-called Nightingale Ward—an oblong structure with windows on both sides, stripped of all unneeded accoutrements—was designed to prevent air stagnation and maximize circulation (fig. 15). This basic prototype spread rapidly through postbellum America and famously crystallized in the one-story pavilion style developed by John Shaw Billings in 1875 for the Johns Hopkins Hospital. With the triumph of the germ theory, such arrangements were deemed neither necessary nor desirable. Hospitals were no longer regarded as dangerous disease quagmires; rather, they were specialist sites where infectious germs were identified, isolated, and dealt with. This transformation meant that the hospital could now promote itself as a scientific shrine with diagnostic laboratories and clinical technologies; and the wealthy willingly came for care. Architecturally, the flat pavilion began to disappear, to be replaced by the towering column. Culturally, the general public met the world of high-tech science when it crossed the threshold of hospital space. The hospital had established itself as the scientific nerve center of the medical world.

But it was not just miasmas and germs that were the objects of medical management within the confines of the hospital; the patients too were subject to various forms of disciplinary regulation. For hospitals have also been moral spaces manifesting the values of their surrounding cultures. Under the patronage of charitable institutions, for example, hospitals expressed their patrons' spiritual ideals. In such environments patients were often treated as "moral minors" in need of correction and instruction. Medical prescription and moral orderliness thus went hand in hand. Rigid rules governed behavior in communal wards. To meet the demands of centralized supervision and training in moral compliance, an austere architectural structure was

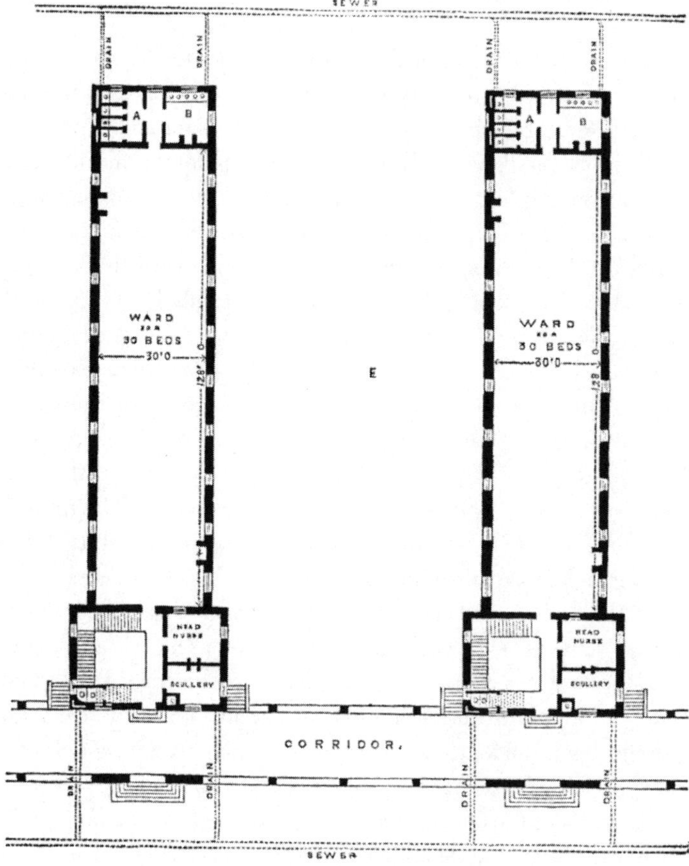

A. Ward Closets.
B. Bath and Lavatory.
C. Lift in Scullery.

D. Private Closet.
E. Ornamental Ground.
Ward Windows to be 4 ft. 8 in. in the clear.

15. Plan of a typical Nightingale ward. The spatial layout reflected Florence Nightingale's miasmatist conviction that circulation was of primary importance in hospitals in order to prevent air stagnation. The requirements of light and ventilation were to be met by one window to every two beds. This spatial arrangement was a radical departure from eighteenth-century practices where the walls directly hindered natural ventilation.

believed necessary to provide patients with both medical and moral therapy. Decorated porticoes and fancy arches may have graced the main entrance, but the patient annexes were plain and unvarying. While the hospital's public facade celebrated its donors' munificence, the patient wings conveyed virtue through orderliness and regularity (fig. 16). Seen in this light, hospital design was a sermon in bricks and mortar on the medical benefits of moral discipline as fundamental to healing. The hospital's walls silently but sternly let its occupants know how doctors and patients were supposed to relate to each other, how visitors should behave, who was in authority, and where medical and moral power resided. And lest the message be missed, a plaque in every ward of St. Thomas's Hospital in London told its pauper sick, "Cleanliness gives Comfort; Sobriety brings Health." In this way medicine played its part in managing the social body by disciplining individual bodies. The aesthetics of hospital space have thus always articulated the core values and beliefs of the medical profession.

If hospitals have been vehicles for transmitting moral values, they are no less sites where new ethical dilemmas have been created. Consider how this is dramatically revealed in the emergency room. And bear in mind that a century ago accident victims were not brought to a hospital at all; they were routinely treated in domestic space, often the kitchen. The concept of a medical "emergency" was not even part of the clinical lexicon a century ago. Nowadays, in the case of cardiac arrest, the patient is frequently rushed to the hospital by a paramedic team, kept technically alive by rhythmic chest compressions, artificial respiration, and a whole repertoire of electronic devices during a speedy transit from the point of attack. In a space dedicated to emergency treatment and equipped with an impressive range of laboratory-style appliances, the physicians on call find themselves facing new ethical problems—generated by the very clinical technology that has delivered the latest case to them—over whether to continue trying to resuscitate. Doing the "right" thing is not just a clinical calculation or a fiscal assessment. It is a moral judgment, and one extraordinarily local to the conditions of its making. Just who the doctor is, is under negotiation; for in encountering a patient who has

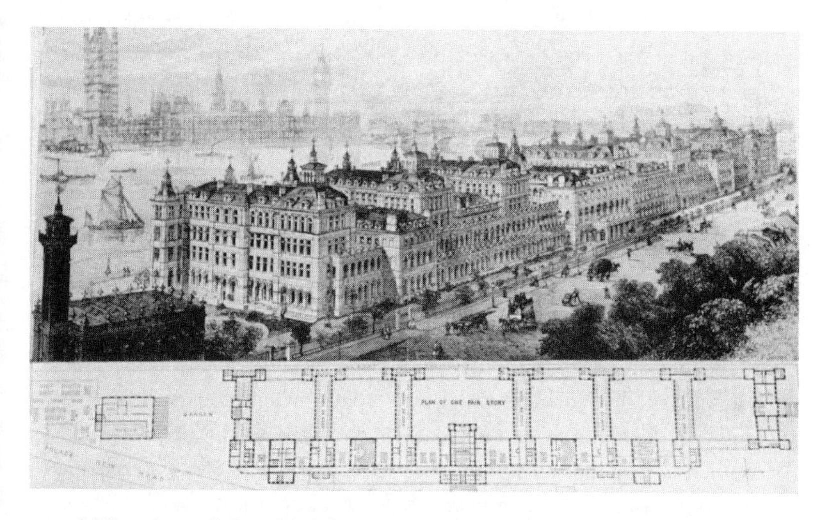

16. Elevation and plan of St. Thomas's Hospital, London. Behind this imposing structure, with its fine neo-Gothic pavilions, were wards that expressed medical and moral orderliness and regularity.

no say in the matter, the physician faces up to his or her self-understanding as a "healer" in the frontier zone between life and death. The space of emergency medicine is a space of ethical practice where unique clinical and moral choices have to be made.

While we now routinely think of hospitals as sites of scientific knowledge and medical education, their acquisition of such functions has a history too. At least in part these initiatives—which surfaced in late Enlightenment Europe—reflected broader changes in natural philosophy as the new medicine found inspiration in the empirical methods championed by the likes of the English philosophers Francis Bacon and John Locke. In eighteenth-century France, for instance, the triumph of experience over theory that Locke's disciple Étienne Bonnot de Condillac advocated was taken up by P. J. G. Cabanis, who preached the value of hospitals for medical instruction and investigation and railed against the sterile speculation of "the old medicine." In Edinburgh John Rutherford began clinical lecturing in the mid-

1700s, and a dedicated teaching ward was established. Besides, the hospital could provide a steady supply of unclaimed cadavers for anatomical instruction in dissection halls. By the middle third of the next century, the practices of bedside diagnosis, medical students' "walking the wards," and daily visits to the morgue had been firmly established. All of these confirmed that medical knowledge was gleaned from local, on-the-spot experience and from deciphering patient symptoms. Not surprisingly, it was out of such experience that the primary importance of standard clinical arts like inspection and percussion, the invention of the stethoscope (by R. T. H. Laennec in 1816), and the use of the compact case history were born. The bedside had emerged as itself a diagnostic space where the student was trained to read the signs of disease. Medical knowledge acquired and applied in hospital settings was thus part and parcel of a wider economy of healing that encompassed technical know-how, hard-won local wisdom, hands-on experience, and moral management.

The idea that hospital interiors are readable cultural spaces is perhaps nowhere more clearly disclosed than in what were called insane or lunatic asylums. Certainly asylums did not exhaust the spaces of insanity. In the medieval period, for instance, the "mad" might be found wandering in fields and forests, restrained in jail-like constructions, sheltering in havens of relief, or hanging around quasi-religious sites of holy waters. Our focus, though, is on the asylum, and here the multilayered character of institutional space dramatically surfaces. To begin with, asylums have regularly been sites of surveillance dominated by the imperatives of supervision and control. This means that, unlike the general hospital ward, their inpatient units have often been small or single rooms, organized to maximize scrutiny. To be sure, detailed architectural arrangements differed from place to place: in Germany, the 1655 plan of Joseph Furttenbach took the form of a Roman cross, the late eighteenth-century French asylum at Salpêtrière had back-to-back cells around a central square, and Glasgow's lunatic asylum erected in 1810 was built using a panoptical cross ward system. But the emphasis was on prisonlike surveillance. As for the internal management of space, the ways inmates have been spatially organized is itself a commentary on the social order. At Salpêtrière during the era of the French Revolution, the layout segregated patients into a bipolar taxonomy of "curable" and "incurable" and reserved enclosures for "idiots," "escapees," and "sowers of discord." Clearly the asylum was as much a space of social policy as of medical treatment.

Asylums were scientific spaces too. Take the fact that until the early nineteenth century the insane were deprived of heat in their cells. This was done on the literally dehumanizing medical principle that since they had lost their reason—the feature that distinguished human beings from the rest of the animal order—they were simply beasts and therefore did not feel the cold. Such designations made it entirely "reasonable" to deploy a frightening arsenal of restraining equipment—leg locks, iron chains, screw gags, restraint chairs, and so on—to exercise medical control. In the Middle Ages the aim of incarceration was exorcism; in the seventeenth century it was reestablishing political order; during the Enlightenment it was disciplining "unreason." Whichever applied, the asylum was a site of scientific and moral therapeutics. Not surprisingly, behind the sometimes palatial exteriors—as with the infamous Bethlehem Hospital (Bedlam), built in London in 1676—was an interior landscape of fear (fig. 17). Indeed, in some institutions "fear therapy" replaced chains as the favored mode of treatment. Francis Willis, famous for his treatment of King George III in the mid-eighteenth century, boasted that he could tame a maniac merely by the mesmeric power of eye control. Here, as in France, Germany, and Italy, was a moral regimen of authority that traded in a compound of rewarding good behavior with favors, instilling terror of chastisement, and gently distracting a patient from delusions. In large measure the idea was that treatment should be aimed at the emotional and intellectual faculties alike.

Asylums have been spaces both of surveillance and of scientific-moral therapy. They have no less been sites of public entertainment, at least until the end of the eighteenth century. At the colloquially christened Bedlam, visitors could pay a penny to walk around the wards for amusement. At times this took the form of active spectator

17. The palatial facade of Bethlehem (Bedlam) Hospital in 1676. Behind this regal frontage was a landscape of fear where most patients were kept behind bars to leave the corridors free and safe for visitors who came for amusement.

sport with, as one observer recalled it in 1753, visitors provoking victims into "furies of rage." Pandemonium would erupt, with inmates clanking chains, hammering on doors, and screaming in anger or frustration or sympathy. The resulting chaos confirmed in the minds of the sane just how profoundly "other" was this space of unreason. The asylum could thus take on the dimensions of a circus precisely because it represented reason's alter ego. In that great "city of reason," Enlightenment Edinburgh, for example, the asylum was located in a small institutional cluster of buildings on the southwest margin of the old town that included a charity workhouse geographically removed from the exalted sites of intellectual brilliance occupied by the literati. The city's very layout thereby exemplified the spaces of reason and unreason, light and darkness. The social exclusion and geographical isolation of people supposedly without reason played a crucial role in the constitution of the Age of Reason itself. Enlightened thinking simultaneously created those benighted spaces and expelled them to the edges of social significance.

By the nineteenth century the hellish associations with which many madhouses were yoked began to be replaced by different imagery—though abuses certainly remained. Psychiatric practitioners like John Conolly and W. A. F. Browne now celebrated the asylum as a progressive institution. Inside, asylums were to be spacious, airy, and elegant, fitted out with galleries and music rooms; outside, they were to be positioned on elevated sites and surrounded by gracious gardens with extensive walks (fig. 18). Exterior geography, it was believed, was as important as interior design. Consider the opinions routinely advertised in the *Asylum Journal of Mental Science*—a serial that came into being in the mid-nineteenth century on the cusp of psychiatry's efforts to both professionalize and medicalize itself. Strenuously protesting that it was physicians who should care for "lunatics," it sought to synthesize medical and moral treatment of the insane. And it was in the context of this "medical-moral discourse" that extensive grounds and gracious gardens were promoted as a means of removing patients from disturbing sensory inputs. In practice this often meant that rural situations were preferred. An antiurban rhetoric predominated because towns were supposed to militate against the required medical-moral regimen. Tranquillity and serenity were the mental conditions sought within the asylum; they were to be diagnostic of its natural landscape too. In seeking to secure such environments, visiting magistrates would evaluate everything from soil type and rock form to water supply and climatic conditions in both moral and medical terms. "A poor, cold, stiff clay," a contributor to the *Asylum Journal* insisted in 1856, "is by no means eligible for the site of a lunatic asylum." Plainly, an asylum's geographical location was as fundamental to the recovery of mental health as its interior spatial arrangements. For as the superintendent of the Coppice lunatic hospital in Nottingham observed, landscape views could divert deranged minds from "imaginary grievances" and "gloomy and distressing thoughts." Here again medical knowledge was located

18. The State Asylum for the Insane at Tuscaloosa. The site of the asylum, amid extensive grounds, reflected John Conolly's principles. A pleasing external environment was considered a key aspect of moral and psychiatric therapy.

within a wider moral order in which connections between interior psychology and external geography were assumed. Moral judgment, mental state, and medical treatment were intricately interwoven.

The Body of Scientific Knowledge

Spaces of therapy are not restricted to architectural structures like the hospital or the asylum. The body itself has often been a site of scientific diagnosis. Edward Jenner, for example, who discovered that dairy workers who had been exposed to cowpox were immune to smallpox, deliberately infected people with cowpox in 1796 to protect them from smallpox. Animal bodies too continue to be spaces of scientific knowledge. It has been estimated that over 100 million animals are used each year in scientific experiments. Whether in laboratories, in the field, on spacecraft, or in a dozen other places, the animal body has been used to test pharmacological, cosmetic, and medical products and devices. Rabbits have been used in toxicology work, rhesus monkeys for experimental surgery, rats in polio research, and horses in investigations of emphysema. And perhaps most spectacular of all, the humble fruit fly was the subject of intensive research in experimental genetics at the University of Columbia's squalid fly lab throughout the first half of the twentieth century.

In this basic sense, the body is a space of scientific endeavor. In many sites—some sinister and secretive, others curative and civic—bodies have been the objects of experimental inquiry. But body space merits our attention in various other ways too. Scientific knowledge has routinely been considered incorporeal and transcendent. And yet, as we will presently see, science has been profoundly embodied in all sorts of ways. Attending to this most local of scales—the body itself—is thus an essential component in cultivating a geography of scientific knowledge and practice.

Because we have just been considering the hospital as a diagnostic space, it will be useful to continue this theme for a little while and dwell on the body as a site of medical experiment. Medical trials, of course, are always executed in some social space, and this is what gives meaning to the practices that are undertaken. Tests carried out on women using oral contraceptives are illustrative. Though the results were routinely presented as demonstrating the effects of the drugs on the "female body," in fact the contexts within which women's bodies became scientific sites made dramatic differences to the very nature of the project. Initially tests were carried out on infertile women in the early 1950s with the aim of inducing pregnancy. The discovery that progesterone inhibited ovulation resulted in larger-scale trials of the pill. The original project was significant because immediately after World War II childlessness was regarded as undesirable. Within a decade mores were changing, and millions of women were turning to oral contraception as a means of planning families for a modern world. Conditions were radically different for the Puerto Rican women who were used in another pilot study in 1955. The long history of colonial population policies that Puerto Rico had already experienced, the fact that contraception was illegal in many states of the

United States in the 1950s, and a fear of possible breeding grounds for communism all made the United States' former colony an ideal space of experimental practice. In this context, contraception was regarded as an essential weapon in a racial struggle for global dominance in a world now believed to be carrying a population time bomb. Here the female body—construed as biologically fecund and ecologically risky—was at once a scientific site and a space of cultural conflict.

The human body, however, has been a site of scientific knowledge in altogether more sinister circumstances. Witness the grotesque medical experiments in what was euphemistically known as racial hygiene in Nazi Germany. In a culture mesmerized by a xenophobic politicizing of biology, the idea that there were "lives not worth living" began to grip. Adult and child euthanasia of the physically and mentally retarded fit comfortably with the campaign to eliminate various ethnic groups. The notorious experiments during the 1940s in such concentration camps as Dachau, Auschwitz, Buchenwald, and Sachsenhausen were merely the extension of an already well established social policy. Here the adjective in the label "Nazi science" certainly makes sense. For in these corrupt spaces victims were compelled to drink sea water, to undergo limb transplants, and to endure temperature extremes to provide scientific knowledge. How long could a person survive in icy water? What effects do particular bacteria have on the body? How effective was a newly developed vaccine for spotted fever? How long does it take a man to die in very low pressure? What happens when a woman is infected with some malignancy and then treated with a new drug? Answering these scientific questions in the dark space of Buchenwald alone cost eight thousand Russian lives. Here the price of scientific light was moral darkness.

But it is not just in the context of medical experiment that bodies have been sites of scientific inquiry. Alexander von Humboldt, Prussian geographer and scientific traveler, used his own body as a recording instrument on his expedition to South America between 1799 and 1804. To be sure, he took with him a seemingly endless supply of standard apparatus—chronometers, sextants, dipping needles, compasses, barometers, thermometers, rain gauges, aeromotors, theodolites, an achromatic telescope, a cyanometer (for measuring the blueness of the sky), and on and on. But instrumentation, we should recall, was intended to extend the range of human sensory organs. The seventeenth-century moral and political philosopher John Locke suspected that angels were blessed with microscopical eyes. Other contemporaries added that the biblical Adam had no need of spectacles, living as he did in a pre-Fall world. Optical devices, it was suggested, could overcome the sensory frailties of fallen humanity and endow natural philosophers with Edenic faculties. Thus Robert Hooke, in his *Micrographia* of 1665, insisted that instruments could help observers come close to Adam's prelapsarian capabilities. So, even with the use of appliances, science was still a profoundly embodied pursuit.

And yet, in an even more immediate sense, Humboldt had to depend on the reliability of his own body to acquire knowledge of the environments through which he passed. When he applied electrodes to himself to ascertain the effects of an electric current on a secretion of blood and serum derived from deliberately raised blisters on his back, he was using his body as itself an instrument. Undeterred by the decidedly uncomfortable results, he later repeated the test, this time using the cavity left after a tooth extraction. And then in South America both he and his companion Aimé Bonpland used their own bodies as virtual Leyden jars to test the discharge from electric eels—with fairly nasty results. In a yet more general sense, bodily changes registered the shifting environmental conditions that both Humboldt and Bonpland experienced as they made their ascent of Mount Chimborazo in June 1802: they got dizzy, their eyes became bloodshot, breathing was difficult, their lips oozed blood. Yet this was precisely to be expected because, as Humboldt himself later reflected, the body was "a kind of gauge" registering atmospheric rarefaction. Barometric readings merely confirmed what their bodies had already told them. For Humboldt these were life-changing experiences. They established his massive reputation as a scientific traveler. But they had such a visceral impact that forever after he kept his dwelling quarters back home heated to tropical temperatures. Whatever effect he had on the tropics was nothing to the effect the tropics had on him.

In acting as an incarnate Leyden jar, Humboldt was not unique. A whole clutch of electrical experimenters during the eighteenth century had freely used their own, and sometimes others', bodies as instruments. In the 1730s, for example, Charles Dufay, intendant of the Jardin du Roi, suspended himself from silk threads to demonstrate how leaf metal was attracted to a body after electrification. He thus went further than an English counterpart who recruited a schoolboy for the same experiment. Besides all this, others subjected themselves to the ecstasies of gas, exposed themselves to blinding sunlight, and used electrical currents to stimulate muscular spasms. In one celebrated case, a seventeenth-century Cambridge graduate allowed fellows of the Royal Society to transfuse sheep's blood into his veins.

When experimenters conducted trials on themselves, they were as often as not engaged in critical exercises in scientific warrant. By this I mean that calling on direct bodily experience was a strategy to provide firsthand testimony that was literally self-witnessing. This seemingly simple appeal to what we could call "the body of evidence," however, was never conducted outside some wider culture of inspection. For a start, in an era of theatrical illusion, public scientific demonstration had to divine ways of avoiding the charge of mendacity or delusion. Moreover there were always questions about precisely whose body—and whose mind—could be trusted to deliver truths. Only the genteel, it was widely believed, were sufficiently self-possessed to bring rational minds to bear on unruly bodies. With too much love of the fabulous and too little love of integrity, "menials," as they were called, could not to be relied on to bear witness. Bodily evidence thus followed the contours of social geography. One mid-eighteenth-century French student of medical electricity, Jean-Antoine Nollet, was unwilling to admit into the circle of experimentation "either Children, Servants, or People of the lower Class." This effectively barred from the "republic of learning" whole sectors of society whose testimony to their own bodies could not be trusted. Self-evidence was always a social product. Appeals to direct perception as the justification of knowledge were thus largely rhetorical. What passed as "immediate experience" was actually the result of negotiations over who was a reliable witness and who had the standing to provide dependable reports.

Josef Mengele's repulsive experiments, Humboldt's embodied collisions with the tropical world, and the electrifying of limbs by eighteenth-century experimentalists serve, in morally different spaces, to remind us that scientific knowledge has taken bodily form in that the body has been the locus of experiment. But we might press a little further in entertaining the thought that scientific rationality has been incarnated in an even more profound sense. Traditionally, reason has been seen as a bodiless thing, incorporeal and transcendent. The processes of thinking, and the products of thought, were considered purely rational undertakings; ideas seemed to float free and clear above the messiness of material existence. In this way knowing could be divorced from living, head work separated from manual labor, minds severed from bodies. The standard breach between embodied life and disembodied knowledge, however, is not all that easy to sustain in the light of several considerations. And here we will pause to consider just three issues that render the rupture troublesome—what I will call self-denial, sex, and situatedness. All point to the inescapably embodied nature of knowledge making.

In earlier times it was widely believed that there was a close connection between bodily discipline—not least dietary temperance—and the capacity for knowledge and wisdom. Brain and belly, spirit and stomach were not divorced in the way they are today. In both the ancient world and the medieval Christian tradition, self-denial was taken to be a precondition for genuine knowing. Asceticism and wisdom routinely went together, as implied by the New Testament record of Christ's fasting and withdrawing into the wilderness to meditate. Insight and indulgence, scholarship and sensuality were incompatible. Rather, bodily subjection and regulated consumption were seen as a prerequisite for achieving enlightenment. In large measure, what motivated these associations was fear of uncontrolled carnality, because gluttony was believed to incite moral mayhem. Temperance was thus venerated as a sacred *and* a cognitive virtue; spiritual health and intellectual clarity alike required bodily moderation. Two ironies are noticeable here. First, the ideal of disembodied knowledge could be achieved only through rigorous management of bodily appetites. One had to be obsessed with the body to escape from it. And second, the very religion that seemed to call for disembodied truth was equally committed to the idea that, in the person of Jesus Christ, truth had taken bodily form. The incarnation was precisely about the word of truth becoming flesh. These ironies notwithstanding, the rigid regimens of disciplined consumption the philosopher was required to engage in bore testimony to the suspicion that genuine knowing was anything but disembodied.

If the kingdom of natural philosophy was forbidden territory to the undisciplined body, it was no more accessible to other bodies—particularly those of women. For all the rhetorical claims to the disembodied character of scientific knowing, there was a long-standing "understanding" that female corporeality rendered women unsuitable for intellectual pursuits in general and for science in particular. Scientific space, by and large, was masculine space. This is not to say that women have never engaged in scientific pursuits. In early modern Europe, for example, noble birth or craft skills did allow some women to participate to some degree in science. But on the whole women were excluded from the domain of natural philosophy, often on corporeal grounds. In the ancient world, given the idea that the body was composed of the four elements of earth, air, fire, and water, the female body was considered inferior because it was supposedly deficient in heat. Much later, in the eighteenth century, the female skeleton was declared to be unfinished and distorted, the configuration of the skull in particular being taken to show intellectual inferiority. Soon evolutionary speculation would lead some to suggest that the female body was in a state of arrested development. And the same line of anatomical argument was used to marginalize non-Western peoples.

Given these preoccupations, it is not surprising that speculations of this stripe were called on to keep science a white, male preserve. On the racial front, such celebrated eighteenth- and nineteenth-century philosophers as Hume, Kant, and Hegel were certainly in sympathy with this viewpoint. Hume insisted that the people of the tropics were "incapable of all the higher attainments of the human mind"; Kant dismissed the peoples of the same zone as cripplingly lethargic; and Hegel insisted that Africans had "not progressed beyond a merely sensuous existence." As for the exclusion of women, it was in the very 1834 publication where he coined the term "scientist" (tellingly, a review of a scientific work by Mary Somerville) that the English astronomer and philosopher of science William Whewell confessed that there was, after all, "a sex in minds." Forty years later Henry Maudsley, professor of medical jurisprudence at University College London and a leading advocate of an evolutionary science of mind, spelled out the wider implications of this stance in an article titled "Sex in Mind and Education." By then, many Victorian scientists were regularly feeling the urge to declare themselves on science and "the woman question." Darwin told the readers of *The Descent of Man* in 1871 that "man is more powerful in body and mind than woman." A few years earlier Thomas Henry Huxley had expressed his disquiet at what he called the "new woman-worship," not least because he was sure that "five sixths of women . . . stop in the doll stage of evolution." And George John Romanes went so far as to claim that, psychologically speaking, males and females belonged to different species. Meanwhile a host of figures felt the need to remark on the childlikeness of women or on woman as "undeveloped man." The late nineteenth- and early twentieth-century American psychologist G. Stanley Hall, for instance, insisted that girls should be educated primarily for motherhood. In the light of these prescriptions, it is not surprising to find observations like that of the American Edward Clarke, author of *Sex in Education* (1873), disparaging female college graduates as "mannish maidens" for whom the price of education was "undeveloped ovaries." It became a common thing to find writers turning to current biological wisdom to justify what they called the physiological division of labor. No less common were the medical warnings issued to women travelers about the heavy toll a tropical climate exacted on the female body and psyche. Women who engaged in field science or had educational aspirations not only were risking their own bodies and their children, they were jeopardizing the race by taking a retrograde evolutionary step.

The female body, then, was long seen as an illegitimate site of scientific learning. Ironically, only certain bodies—male and white—had the capacity to generate disembodied knowledge. And that "disembodied knowledge" included the idea that mental differences between the sexes were biologically based! Women had become the victims of the pursuit from which they were by and large barred—science itself. The consequences were certainly long lasting, not least on the institutional geography of science. It was only in 1945 that women were admitted to the Royal Society; and it was not until over thirty years later—1979—that the French Académie des Sciences opened its doors to women.

The rigors of self-discipline as a precondition of readiness to receive wisdom, and the edging of women's bodies to the margins of scientific pursuits, suggest that, despite robust protestations to the contrary, there has been a persistent suspicion that scientific knowledge is all too embodied. And for good reason. If it is true that scientific instruments help knowers "sense" the world in ever more subtle and sophisticated ways, then implements can be considered extensions of sense organs. The twentieth-century chemist and philosopher Michael Polanyi fastened on this feature of scientific instrumentality when he judged that using instruments enlarges our senses. In one way or another, then, the body is always in service as the "basic instrument" of our intellectual engagement with the world. "Every time we assimilate a tool to our body," Polanyi wrote in 1959, "our identity undergoes some change; our person expands into new modes of being." Accordingly, "in all our mental achievements we rely ultimately on the machinery of our body." And there are additional inferences to be drawn from the necessarily embodied nature of knowing. Given that bodies are resolutely located in space, there are grounds for suspecting that scientific knowledge is always positioned knowledge, rationality always situated rationality, inquiry always local inquiry. The physicality of human bodies and the artifacts they employ mean that the knowledge humans produce is inescapably partial. It constitutes a view from some particular location. On this account, science displays rather than transcends human particularity—in terms of race, gender, class, and in all likelihood a host of other factors. The aggregate judgment of Victorian biologists on the intellectual ineptitude of women certainly seems to support this suggestion. Because the body is a site of science, scientific understanding is always a view from somewhere. It is always local knowledge. After all, whether science is practiced in a laboratory, a museum, a botanical garden, a field station, a hospital or wherever, these spaces are always occupied by embodied investigators.

Of Other Spaces

The sites we have visited so far certainly do not exhaust the spots where scientific knowledge has been generated. Laboratories, museums, hospitals, and so on are conspicuous landmarks in the landscape of scientific endeavor. But other locations have been important too. Cathedrals are a case in point. In medieval times the church's need to ascertain when Easter would fall each year was easily determined in theory. It was the Sunday after the first full moon after the vernal equinox. But it was extraordinarily tricky to figure out in practice. Because the time of the sun's return to the same equinox was a key feature of the calculation, one preferred means of addressing the problem was by laying out a meridian line from south to north in a darkened building and observing the shifting position of the sun's noon image on different days as it shone through a hole high up in the structure. From the late Middle Ages right up into the eighteenth century, cathedrals were used for this purpose. They were thus key sites of astronomical observation and remarkable accompanying mathematical computations. Ironically, in the very heart of the Papal States, the oldest of these cathedral observatories recorded measurements that called into question such standard dogmas as the doctrine that celestial motion was perfectly circular.

Other sites of scientific inquiry have been a good deal less stable.

Consider the ships used during voyages of scientific discovery. Many of these have achieved near mythic status in the annals of science—La Pérouse's *Astrolabe,* Cook's *Endeavour,* Darwin's *Beagle,* Huxley's *Rattlesnake,* Wyville Thomson's *Challenger.* Not only did these carry scientific instruments, they frequently *were* scientific instruments in their own right. James Cook's charting of New Zealand is illustrative, on board a ship that incidentally carried botanical equipment, artists, and French-horn players! For it was through his tracking of the *Endeavour*'s geodetic position that Cook inferred the contour of the coastline. In this way the ship became a surveying instrument that delivered the lineaments of the coastal fringe without ever touching it. The very computations that permitted Cook to set his course delivered him a cartographic shadow of the coast his vessel left behind. In its capacity as a surveying device, as in its housing instrumental gadgetry, the ship has been an important site of science.

Another mobile site of scientific inquiry is the tent. Functioning as a transitory workshop, it has bridged the gap between the laboratory and the field. But it has performed other roles too, not least in anthropology, where "getting under canvas" became a rite of passage that both conferred professional status on the initiate and established ethnographic authority. Moreover, in at least some places—mid-twentieth-century Rhodesia, for example—tent dwelling enabled anthropologists to trade on the standing already enjoyed by colonial officials and government surveyors. Political and scientific authority were mutually reinforcing. Yet in the long run the association was counterproductive. Emerging nationalism raised suspicions that anthropologists were government spies, so new techniques of fieldwork, compatible with decolonization, began to be devised. Since the tent was associated with colonial officialdom, anthropologists who wanted to distance themselves from government administrators began to use trailers and campers for rural fieldwork because they were seen as more politically neutral.

The ship and the tent are constituted as elite spaces of scientific practice because of the activities carried out by their temporary inhabitants—scientific surveyors and specialist ethnographers. But other elite arenas, less professional though more aristocratic, have been sites of scientific knowledge too, not least the royal court. In the late sixteenth century, Galileo's performances in the cause of the Copernican theory before the nobility of his day, for example, conformed to the accepted modes of communication that were embedded in the courtly culture of early modern Italy. As an expressive space no less than a civic one, the court maintained chivalric codes that set limits on how Galileo could expound his views in this princely setting. The court has also been a performative space where understandings of the natural order were theatrically enacted. During the first half of the seventeenth century, court masques in Britain were often the vehicle for declaring the unity of the British Empire and its supposedly special destiny. This species of political theatrics routinely resorted to geographical factors to guarantee national identity under the wise authority of the monarch. In this way the court became an arena in which natural knowledge—of woods, mountains, ancient ruins, and so on—was mobilized to justify political order.

Elite spaces have not had an exclusive monopoly over scientific endeavor. A range of public places have also played host to the production and dissemination of scientific knowledge. Such sites are perhaps less visible to the scholarly eye, though, in view of their role in popular culture and a long-standing sense that they could be, at most, venues for popularizing science. But the boundary line between philosophical gentlemen and what has been colorfully called "gimcracking virtuosos" is harder to draw than one might imagine. For one thing, many eighteenth- and nineteenth-century "popularizers" were serious experimenters themselves—the chemist Humphry Davy, the pioneering student of electromagnetism Michael Faraday, the quarryman-geologist Hugh Miller, and Darwin's "bulldog," Thomas Henry Huxley. Any attempt at rigid demarcation here is therefore likely to be misplaced. What is significant is the way science washed up in what we might now think of as unlikely public places and was connected with social classes markedly different from aristocratic patrons and professional elites. Two instances will serve as illustration—the coffeehouse and the public house. In each case, even though we are entering spaces accessible to the public, our finger is on the pulse of a different segment of the social body.

The coffeehouse was a key site, along with the lecture theater and salon, in the making of what has been called "the public sphere"—that realm of social interaction vital to the emergence of critical sociability, rational dialogue, and the exchange of information. Fundamentally a place of bourgeois encounter and thus central to the genesis of commercial capitalism, the eighteenth-century coffeehouse was a space created for the public use of reason, chiefly through the medium of newsprint. These spaces were protopolitical institutions. But they were also sites for promoting Restoration science. London coffeehouses, for example, hosted scientific lectures and experimental displays and thereby bridged the gap between early entrepreneurs and natural philosophers from Gresham College and the Royal Society. The seventeenth-century experimentalist Robert Hooke regularly visited the London coffeehouses, where he engaged in scientific discussion with Robert Boyle, Henry Oldenburg, and other key figures from the Royal Society. And in Plymouth, the local coffeehouse sponsored a debate in the 1680s on whether brain wounds were curable. Given these activities, it is not surprising that the coffeehouse was sometimes dubbed the citizens' academy—a popular university in which class divisions were broken down and useful knowledge was propagated. For these very reasons the institution was, from time to time, suspect among those who thought it hostile to both tradition and monarchy. Coffeehouses, one critic remarked, made every porter into a statesmen and were hotbeds of cultivated sedition. Whichever is the case, as George Steiner has recently noted, the coffeehouse defines "a very peculiar historical space"—"of discourse, of shared leisure, of shared exchange of disagreements."

The public house conjures up an entirely different social atmosphere. For one thing, coffeehouses admitted women, and their patrons thus soon attracted the charge of effeminacy. Moreover, they were seen to be in contention with the traditional sports of the tavern, such as cockfighting. A marked cultural difference thus characterized these two public spaces from the start and became more pronounced as time went by. In the twenty years between 1830 and 1850 the English village inn, in which all classes had eaten and drunk together, was transformed into an exclusively working-class space. And this was the environment par excellence in which artisan botany was practiced in early Victorian England. Science carried on in this location bore highly distinctive marks. Artisan botanists would congregate in public houses on Sunday mornings for botanical meetings where they shared plant knowledge, exchanged specimens, and consulted botanical textbooks. The practical nature of this interchange was paramount, as befitted florists, gardeners, and herbalists, though many participants displayed a thorough command of Linnaean taxonomy. Joined by a love of plants, these enthusiasts created botanical societies within public houses and pooled hard-earned cash to purchase horticultural books and create herbaria that were looked after by the innkeeper. In such spaces, a renegade scientific community, mutually self-rewarding and smitten with collectors' contagion, became sufficiently expert for gentlemen like J. D. Hooker of Kew Gardens to resort to them for specimens and skills alike. Science in the public house thereby challenged the long-standing opposition between head and hand, between philosopher and craftsman. It also reopened scientific discourse to popular interest groups long denied access to elite spaces of scientific inquiry like the laboratory. Seen in this light, the public house was a cultural space that contested the dominant scientific regime of the time.

By reminding ourselves that science has been part of the public sphere and has been practiced in a variety of popular arenas, we considerably widen our awareness of the range of spaces in which scientific knowledge has been produced and propagated. Doubtless the list could be elaborated in extenso, for science has been conducted or communicated in one way or another in libraries, lecture theaters, salons, nurseries, observatories, churches, workshops, artists' studios, mechanics' institutes, learned societies, stock farms, shipyards, game reserves, and on and on. What all these spaces share, both popular and elite, is that—in common with all other places—they are made. They become what they are through the activities that "take place" in them and the human practices that constitute them. In turn these arenas are active in producing the kinds of subjects humans are in those spaces. Space is therefore not dead, inert, and fixed; rather, it is lively, shifting, fluid. Space is animated by events. It is always a production. And scientific space is no exception.

* * *

The enterprise we casually refer to as "science" embraces a huge range of activities carried out in many venues. In these miscellaneous spaces, nature has been differently experienced, objects have been differently regarded, claims to knowledge have been adjudicated in different ways. It is only when the practices and procedures that are mobilized to generate knowledge are located—sited—that scientific inquiry can be made intelligible as a human undertaking. In important ways, scientific knowledge is always the product of specific spaces. To claim otherwise is to displace science from the culture of which it is so profoundly a part.

Region

CULTURES OF SCIENCE

Global forces are acting to homogenize our world. Yet we still live on a highly differentiated planet. Topographically, climatically, politically, culturally, and commercially, our world is divided into a sequence of regional mosaics. But these regions are not simple, straightforward segments of earth space. The commonplace, if colorful, maps of the great natural regions of the earth that routinely crop up in the early pages of our atlases need to be recast in at least two ways if we are to catch a glimpse of the significance of regionalism for the practice of science.

First, regional difference cannot be reduced simply to facets of physical nature or the observable components of material culture. Every place, it was observed nearly a century ago, has "its *genius loci,* of which the poet is usually the best interpreter"; every region has its own distinctive "regional psychology." Even if such professions are a touch too mystical, there is no doubt that traditions of thought, channels of intellectual exchange, linguistic heritage, educational customs, codes of cultural communication, forms of religious belief, and numerous other constituents of human consciousness are decisively operative in producing regional identity. As a medium for the expression of human culture and as an element in the shaping of social life, the region plays a crucial role in making a society's sense of selfhood. Second, even when we enlarge the concept of region to include the "geographies of the mind," we should not consider these as fixed, static entities. Regions are not hermetically sealed "givens." They are better thought of as outcomes, the products of forces both within and beyond their contingent boundaries. The global dynamics of economic transformation have delivered a world characterized by uneven development and social diversity. Regions are thus constructed through the tangled circuits of social relations that, at different scales of operation, produce and reproduce local senses of place, power, and personality.

In the light of these considerations, we can begin to anticipate some of the ways different regional settings may influence both the conduct and content of scientific endeavor. Everything from styles of patronage, pedagogic traditions, and conduits of intellectual transmission to networks of communication, patterns of social organization, and expressions of religious devotion has conditioned local practices of scientific inquiry and the reception of scientific knowledge. Such regional features, moreover, are not to be thought of as simply "external" to scientific inquiry, as merely the context within which "universal science" is carried out. To the contrary. They have profoundly influenced the doing of science in particular regional environments and the knowledge claims that practitioners have made. Explanations that students of nature advanced reflected the interests of their patrons. They were conditioned by the range of accounts that their faith would allow. They were constrained by the ideological uses scientific theories could be put to. They were molded by the prevailing intellectual culture and the systems through which it was sustained. So a regional geography of science has much to tell us about how scientific knowledge is constructed, about what pass as acceptable ways of getting at reality, and about how scientific claims are justified and stabilized. Scientific credibility thus is not to be thought of as something that is obvious or self-evident. In different regional settings, warrant and trustworthiness have been achieved in different ways. Besides all this, regional traditions have acted to promote or impede technical or theoretical innovation. Regional cultures have appropriated scientific knowledge differently according to their sense of self-understanding and put it to different uses. The very meaning of a particular scientific theory or text has shifted from one place to another. Indeed, scientific inquiry itself has signified different things in different regional environments.

In all these ways, and at every scale of analysis from the continental to the provincial, science has been marked by regional particularity. It therefore makes sense to locate "science" according to spatiotemporal coordinates: we can coherently speak of Chinese science under the Sung emperors, Arabic science under the patronage of the Abassid caliph al-Mansur, American science in the age of Jackson, or French science in the late Enlightenment. Equally, we can plausibly refer to "Edinburgh science" in Enlightenment Scotland, "London science" in the early Victorian period, or "Charleston science" in antebellum America.

In chapter 2 we visited some of the specific sites of knowledge making. In what follows we move to the regional level to ascertain the significance of this scale of geographical operations on scientific inquiry. Of course this should not be taken to imply that science has not displayed significantly international, transregional features. The periodic table of elements is the same for scientists in London, Lima, and Lisbon. Besides, the very existence of such cultural merchandise as the Nobel Prize plainly attests to some shared criteria of excellence. But internationalism in science, insofar as it really does exist, must be considered a social achievement, not the inevitable consequence of some inherent scientific essence. It has had to be worked at. And the deep political rivalries that have characterized human history mean that even in the past half century, organizational networks and international associations have had to be cultivated in the attempt to transcend the bipolarities of, say, Cold War rivalry. Such ventures continue the great nineteenth-century encyclopedic exhibitions and congresses designed to counter the increasing fragmentation of knowledge. And yet efforts to establish an internationalist scientific credo have proved to be more an aspiration than an attainment. Political antagonisms, national hostilities, commercial competition, and military interests are just a very few components of the Realpolitik that has muted the optimism of those promoting scientific universalism.

Our quarry in this chapter, then, is science in its regional expression. Just how regional and subregional factors have conditioned the production and consumption of scientific knowledge, the way it was received in different places, and how science has expressed or channeled local loyalties will therefore be chief among the matters that now attract our attention. In chapter 4 we will turn to the ways science moves from region to region and to how fundamentally local knowledge has taken on the appearance of universality.

Region, Revolution, and the Rise of Scientific Europe

The idea that Europe was the cradle of modern science has long been a vital ingredient in the West's perception of its own cultural identity. In one way or another, "the European Scientific Revolution" has been depicted as a prominent feature on the intellectual landscape of Europe's history. Some have portrayed it as the greatest revolution in human consciousness since the advent of Christianity. Others see it as the radical casting off of a cramping orthodoxy. Still others characterize it as the decisive triumph of firsthand experience over Scholastic authority. Three "singularities," however, are troublesome here. First, the idea that that there was some single event called "the" Scientific Revolution is the product of self-conscious labeling on the part of apologists and historians. In fact, what came to be gathered into the designation was a whole suite of practices and procedures geared to understanding and manipulating the natural, and indeed the social, world. Second, the idea of a momentous "revolution" suddenly inaugurating modernity fails to do justice to the lengthy historical transformations connecting the medieval with the modern. And then the casual geographical qualifier—the "European" Scientific Revolution—conceals as much as it reveals. That imagined regional unity—Europe—may usefully be prised open to disclose external influence and internal variation.

The idea of an autonomous European science is sustainable only at the expense of a series of strategic exclusions. European debts to Chinese and Islamic scientific developments need to be registered. There were, for example, the influence of Chinese alchemy on European medicine and the astronomical significance of Islamic geodetic methods of determining "the sacred direction" for the purposes of daily prayer. Baghdad played an important part in cultural transmission through the translation and diffusion of Greek medical and scientific works. The mathematical writings of Archimedes, the astronomical and geographical treatises of Ptolemy, and various Aristotelian philosophical texts in translation all spread west from Baghdad to Córdoba. Add to these the influence of medieval Arabic mathematics and al-Biruni's modifications to Aristotle's physics. All seriously compromise the conception of a self-contained European scientific tradition and confirm the suspicion that the idea of Europe as an independent regional entity is itself a gigantic act of geographical imagining.

But the supposition of a seamless European Scientific Revolution is bothersome in another way too. It fails to take seriously enough the regional geography of science. When he crossed the English Channel, Voltaire sensed that he had entered a different intellectual world. All that was solid in Paris melted into air in London. "A Frenchman arriving in London," he wrote in his *Lettres philosophiques* of 1734, "finds things very different, in natural science as in everything else. He has left the world full, he finds it empty. In Paris they see the universe as composed of vortices of subtle matter, in London they see nothing of the kind. For us it is the pressure of the moon that causes the tides of the sea; for the English it is the sea that gravitates toward the moon. . . . Furthermore, you will note that the sun, which in France doesn't come into the picture at all, here plays its fair share. . . . In Paris you see the earth shaped like a melon, in London it is flattened on two sides. . . . The very essence of things has totally changed." Voltaire's experience bore ample testimony to the potency of the quip made by sixteenth-century French essayist Michel de Montaigne, that truth on one side of a mountain was typically considered falsehood on the other.

Voltaire's shrewd remarks imply that particular cultural circumstances in different national settings influenced scientific ideas in markedly different ways. In France, the shape of political authority under the Sun King directly affected scientific endeavor. In England the advent of the Civil War and accepted codes of gentlemanly conduct had important roles in the acquisition of scientific knowledge and in the ways cognitive disputes were resolved. For the German-speaking lands, religious educational institutions were decisive. Elsewhere circumstances were different, of course, and sufficiently so to make entirely plausible the idea that the rise of "scientific Europe" had a geography as well as a history.

By turning now to conditions in several parts of sixteenth- and seventeenth-century Europe, we can begin to flesh out some of the ways early modern science followed the contours of geocultural variation. What becomes clear is that different kinds of enterprise contributed in different places to what came to be known as science; that scientific knowledge in these regional settings was intimately connected with religious and political affairs; and that it makes sense to attach regional adjectives to the scientific inquiries undertaken at the time. Comprehensive survey, of course, is not my goal here. What follows is meant to be suggestive rather than exhaustive. By dwelling on Italian science, Iberian science, and English science in the period, we can begin to disclose something of the difference that regional circumstances make to scientific pursuits. For other regional settings, other stories must be told.

By 1500 the Italian peninsula was one of the most highly urbanized zones on earth, with its "hundred cities" including such cultural centers as Palermo, Milan, and Venice. It also enjoyed a long history of banking, a culture of private schooling and book collecting, and since the completion of Dante's *Divine Comedy* in the early fourteenth century, an increasingly common literary language in the form of Tuscan Italian. Situated at the hub of the Renaissance revival of classical learning and with such distinguished universities as Bologna and Padua, Italy was remarkably influential in the early flourishing of science in Europe. In consequence, by 1600 treatises—both ancient and modern—in astronomy, ballistics, geography, mathematics, and mechanics were circulating in Italian rather than Latin. At the same time, the establishment of institutions for maintaining doctrinal orthodoxy—such as the Society of Jesus (1540), the Index of Prohibited Books (1543), and the Council of Trent (1545)—made Italy a precarious environment for certain scientific pursuits.

In such circumstances princely patronage was vital to the cultivation of scientific inquiry, not least because of its potential technical applications. Sometimes for commercial purposes, though just as often for reasons of prestige and self-promotion, leading families like the Medici—who had been prominent merchants and leaders in Florentine politics since the thirteenth century—invested in science as cultural capital. Any would-be natural philosopher in search of a patron was well advised to find ways of offering some scientific gift that would bring glory to baroque rulers who were obsessed with image and status. This was particularly so for those engaged in new scientific undertakings that did not enjoy the prestige of more established pursuits. Astronomy fell into this category. It lacked the long-standing authority of philosophy in the kingdom of learning. Because credibility in intellectual matters depended not just on cognitive or disciplinary prowess but on standing with the ruling nobility, the making of knowledge was intimately bound up with social affairs. Of course competence and capability per se were not unimportant. Princes were not in the game of patronizing the mediocre. But what passed as intellectual excellence was constituted through a complex interaction between personal ability and princely benefaction. Skills in calculation, observational proficiency, and theoretical insight were not in themselves sufficient to deliver legitimacy in the knowledge-making business. What counted was courtly status and esteem.

So when Galileo managed to acquire the munificence of the

Medici family through a shrewd decision to christen the satellites of Jupiter "the Medicean stars," he moved *himself* from the Venetian University of Padua to the court of the grand duke of Tuscany and his *astronomy* up the ladder of intellectual status. Securing patronage for mathematical pursuits was itself a remarkable achievement. But it was to have unforeseen consequences. Shifting to become "mathematician and philosopher" to the grand duke in 1610 meant simultaneously returning to a state where the influence of Rome was considerably greater. And any newfangled challenges to Aristotelian orthodoxy were likely to attract the attention of watchful pontifical eyes.

The courtly culture of seventeenth-century Italy was not simply a significant regional context in which scientific endeavor was carried out. Accepted modes of disputation at the Italian court directly influenced the way scientific activities were undertaken. In the case of Galileo, the ramifications were profound. The theatrical style in which controversial subjects were handled allowed him to engage in debate with a pugnacity that would have been seen as improper in England. There, drudgery in scientific pursuits was seen as a virtue, the sensational as vanity. But courtly ostentation could carry a large price tag. In the short term it cost Galileo the very papal legitimacy he so coveted. Copernicanism was condemned in 1616, and Galileo was notoriously brought to trial in 1633. In the long term, it contributed to Italy's decline in intellectual liberty (fig. 19). In trying to make sense of what has become a classic instance of the so-called warfare between science and religion, the particularities of regional culture turn out to be of signal importance. Galileo's clash with the church is not to be thought of as an inevitable confrontation between science and theology; rather, it was an embodied struggle between religious authorities and new ways of knowing in a specific regional setting.

For all its notoriety, the Galileo episode was far from representative of all Italian science in the period. The Jesuits, for example, pursued observational astronomy, and fields of inquiry like electricity, medicine, hydraulics, and natural history, which remained untouched by Copernican squabbles, continued to be developed. Such exploits

19. Frontispiece of Giovanni Battista Riccioli's Almagestum Novum, *published in Bologna in 1651. The illustration shows the lightness of the Copernican system compared with Tycho Brahe's theory when weighed in the balance. Works of this sort enabled Aristotelian science to be perpetuated in the papal states during the seventeenth century. In Bologna in particular Jesuits continued to practice observational astronomy, but they avoided dealing with issues connected with the Copernican theory.*

owed much to the virtues of utility and industry that the Jesuits prized so highly. Nor did the distinctive cast of courtly patronage integrate scientific endeavor throughout the Italian peninsula. Italy remained politically fragmented, and conditions in Florence, Rome, Naples, and Venice were different. Each sustained different relations with the papacy. Some subregions were dominated by merchants, others by priests, yet others by bandits. Given these different regional constituencies, it is not surprising that speculative theories were differently received. In Rome, Aristotelian orthodoxy and anxieties induced by the Reformation over the interpretation of scripture persisted. In Venice, the harboring of heretics at the university made it a prominent landmark in the uneven geography of European intellectual freedom.

Other new Italian sites of knowledge were also important in the shifting status of scientific endeavor. Prominent among them were the anatomy theaters that emerged in late sixteenth-century Padua, Pisa, and Bologna, where public dissections of cadavers were carried out, usually at the time of carnival. Here the defilement ordinarily associated with dead bodies was sanctified by having its social meaning inverted. What was criminal outside became science inside. What was profane was made sacred. In the anatomy theater the unseemly achieved legitimacy. Thereby medical science began to find favor in circles that had previously despised it.

In significant ways, then, regional particularities impressed themselves on the form and content of Italian scientific inquiry. Cultural conditions and knowledge-making enterprises were correspondingly different along Europe's western fringe, the Iberian peninsula. Here geographical location itself mattered a good deal. Proximity to North Africa was crucial. From there the tentacles of Arabic culture snaked their way across the entire peninsula. In Spain this influence manifested itself early on in astronomical treatises—like the *Alfonsine Tables*—produced in the thirteenth century by Arabic scholars and made available in new versions for more than three hundred years. In Spanish medicine, too, Islamic influences were so powerful that when a virulent anti-Mohammedanism gripped the

Christian West after the fall of Granada in 1492, only medical treatises escaped the bonfires into which most Arabic texts were tossed.

This strong Arabic presence, however, was not the only distinguishing feature of early Iberian science. The maritime imperatives of Europe's Atlantic margins stimulated a scientific tradition conspicuously different from that of Italy's courtly culture. This trajectory has often been traced back to the shadowy figure of Portugal's Prince Henry the Navigator, who reportedly established in the early fifteenth century a cartographic and navigational academy at Sagres near Cape St. Vincent. There is little evidence to substantiate that romantic story, but naval pursuits were nonetheless critical. In the forty years from 1481 to 1521, Portuguese monarchs took an active interest in nautical matters for imperial purposes. Of course stimulus is one thing, substance another. And no group did more to advance the cognitive side of navigational science than the Jews. Judah Cresques—from the Cresques family that had earlier made Majorca the cartographic capital of the world—was brought to Portugal for this very purpose. Later, after the expulsion of Jews from Spain in 1492, one of the most skilled astronomers arrived in the person of Abraham Zacut. His Hebrew treatise on astronomy was translated into Portuguese, and he played a major part in planning and equipping Vasco da Gama's expedition to India. The Jewish welcome, however, was short-lived as Manuel I forced Jews either to leave Portugal or to convert to Christianity. Many left. But others stayed as *conversos* and made major scientific contributions, not least in medicine.

Iberian science, erected on the tradition of exploration, was stamped by imperial utility. Cartographic expertise and instrumental innovation were at a premium. But the scientific significance of overseas voyaging spiraled well beyond navigational science. Advances were made in the study of terrestrial magnetism and hydrography. Medicinal botany was developed by the sixteenth-century Spanish Jew Garcia d'Orta, who provided original descriptions of a host of Asian plants, such as the mango, cocoa, and camphor. Francisco Hernández's expedition to New Spain (Mexico) delivered specimens and seeds by the sackload (fig. 20), and José de Acosta's *Natural and*

20. An engraving of a New World medicinal plant made from an illustration by the court physician and explorer Francisco Hernández. The advancement of medicinal botany in Iberia crucially depended on the specimens gathered by explorers in the New World and on the information that figures like Hernández collected from local herbalists.

Moral History of the Indies, which made its appearance in Seville in 1590, provided reports of such strange beasts as iguanas. At the same time, because Lisbon and Seville found themselves at a global crossroads in the exchange of intercontinental commodities like cloves and spices, there developed what might be called imperial arithmetic. Mathematical texts, like the one Gaspar Nicolas printed in 1519, dwelt on how to determine the levy on merchandise, how to convert different currencies, and how to deal with variations in weights and measures. But perhaps most of all, direct observation of the faraway began to subvert the authority of the ancients on such matters as the nature of the tropics and the range of plant and animal species on the globe. In 1532, one writer could castigate the geographical ignorance of such authorities as Strabo and Ptolemy. In these and many other ways, Iberian science bore the unmistakable marks of the expeditionary "far side."

This early flowering of Iberian scientific enthusiasm was to fade as the sixteenth century wore on and more and more works of scientific scholarship were placed on the Index of the Inquisition. The story here is complex and intricate, but it is likely that anti-Semitism, the expulsion of *conversos,* and shifts in educational policy under the Jesuits all contributed. Still, Iberian science, stemming in large part from the imperatives of empire, was a markedly different pursuit from that practiced in the Italian court under the patronage of powerful family dynasties. There courtly status was all-important in attaining credibility. In Spain and Portugal it was proficiency in the practicalities of the haven-finding arts and in the crafts of healing that delivered cognitive authority. Scientific inquiry in the Italian and Iberian peninsulas meant very different things—in what was investigated, who had the power to make knowledge, and why certain lines of inquiry were pursued.

Precisely the same was true of England. By the late seventeenth century it had emerged from relative obscurity to become a major player on the field of European science. The catalog of achievements is remarkable. Thomas Digges was advertising Copernicanism by 1576. By the dawn of the new century, William Gilbert's work on magnetism had appeared, as had Edward Wright's notable application of mathematics to nautical cartography. William Harvey's demonstration of the circulation of the blood was made available in 1618, and over the next decade he continued his experimental investigations into animal anatomy. Conceptually, Francis Bacon's insistence on inductive inquiry—the patient gathering of facts—as the necessary first step to the elucidation of general principles in nature came out in his *Novum Organum* of 1620. Intended as a reform of all human knowledge, it constituted an eloquent apologia for erecting science on the sure foundation of method. And it had the added advantage, as he put it in the sixty-first aphorism, of placing "all wits and understandings nearly on a level." By the 1660s, Robert Boyle was conducting experiments on the vacuum using the air pump and applying mechanical theories to chemical phenomena that cast doubt on Aristotle's doctrine of the elements. And supremely, there were the accomplishments of Isaac Newton in the late seventeenth and early eighteenth centuries. His account of universal gravitation and planetary motion, his investigation of optics, his creation of the calculus, and much else besides are all landmarks in the cultivation of modern science.

As happened with Portugal, overseas voyages contributed to this remarkable transformation in regional consciousness. Gilbert hoped that his study of magnetic behavior would improve sailing methods. Bacon, conscious of the empirical riches delivered by seafarers, thought it would be shameful if opening up the *material* globe did not lead his contemporaries to transcend the ancient bounds of the *intellectual* globe. So English science certainly reflected navigational concerns. But what made it conspicuously different from its Italian and Iberian counterparts were the circumstances of post-Reformation Europe's political and religious geography.

In England, the triumph of experimental philosophy—the idea that it was through experiment that nature was best understood—and scientific projects more generally, took place in the midst of religious turmoil. Religious differences lay at the headwaters of the English Revolution, and Protestantism in its various incarnations gained the ascendancy after the mid-sixteenth century. These religious and political currents had a direct bearing on the culture of English science. This does not mean that English science was rooted in some highly particular suite of theological or denominational convictions. Rather, Protestant impulses in England influenced the scientific endeavors of the natural philosophers in a variety of ways.

Take the matter of authority. The more radical Protestants were deeply averse to the kinds of ecclesiastical control exercised by the Catholic Church, and they emphasized the supreme value of personal religious experience—"experimental" religion, as they tellingly called it. Such convictions nurtured an antiauthoritarian stance in matters of natural knowledge. The authority of the ancients and the Aristotelian strictures under which astronomers in Catholic Italy labored were rejected outright by many English Protestants. Spokesmen with Puritan sympathies did not hesitate to castigate what one called "the rotten and ruinous fabric of Aristotle and Ptolemy." In these circumstances scientific explanations did not need to be molded to fit an Aristotelian template. Important too was the practical cast of mind that distinguished some branches of Protestantism. The virtues of hard work, an inclination toward social improvement, and dedication to a life of personal piety fostered a philosophy of self-reliance in harmony with the utilitarian thrust of new scientific enterprises. Such sentiments manifested themselves at the Puritan-dominated Gresham College, where technological applications to navigation and trade held pride of place in the curriculum (fig. 21). And Protestant expectations of the imminent return of Christ and the ushering in of his millennial kingdom fostered misgivings about abstract, speculative disputes of precisely the sort that typified French thought in the period. Theoretical preoccupations were simply a distraction from the more important, if mundane, toil needed to re-create earthly paradise. Besides all this, English advocates of a reformed natural knowledge often used their inquiries as a resource for denouncing what they saw as papal fantasies, clerical tales about the powers of sacred shrines, hermits' fables, and the like. Such myths and legends simply did not measure up to the rigor of Bacon's new method. Indeed in Bacon's own hands,

21. The Puritan-dominated Gresham College was founded in 1598 for instruction in divinity, astronomy, music, and geometry. Largely under the influence of the Calvinist Henry Briggs, a professor of mathematics who also taught astronomy, navigation, and geography, Gresham College became a key location for the study of practical mathematics and experimental natural philosophy. In contrast to university education, it emphasized skills for solving nautical problems and contributing to trade. This particular image, with open fields to the rear, conveys the sense of a Protestant cloister.

natural philosophy was mobilized in the cause of excising the idolatrous from Christianity, of filtering out the fabulous, and leaving behind a residue of Protestant purity. To read the book of nature properly would purge Christianity of those superstitious accretions that Catholicism still clung to.

Given these preoccupations, the flourishing of physicotheology, as it was called—interrogating nature for evidences of God's design—in seventeenth- and eighteenth-century English science is understandable. The character of God was to be found in the orderliness of his creation. Natural philosophers from Boyle to Newton consistently used their investigations to disclose the regularity that the Creator had built into the fabric of the universe and to demonstrate the ways he intervened to preserve its stability. Newton, for example, though hardly orthodox in his own theology, declared in the *Opticks* (1717) that "the main business" of natural philosophy was to establish firm grounds for belief in God. In his will, Robert Boyle made provision for the delivery of a set of annual lectures—the Boyle lectures—intended to confute infidelity, atheism, and deism. Given from 1692 to 1714 and published through his financial largesse, they demonstrated how various forms of scientific endeavor could, in the form of natural theology, act as handmaiden to the Christian religion. The very fact that Boyle felt the need to make such provision, of course, hints at the deistic tendencies that natural philosophy harbored and that flourished as the eighteenth century wore on.

In England, the way many natural philosophers interpreted nature was intended to support Christian theism. But how they read the book of nature was itself a consequence of the revolution in textual interpretation that the Reformation had inaugurated. Allegorical ways of construing texts declined and were replaced by a more literal and historical exegesis of the Bible. These moves bore on how the text of nature was read. Protestant biblicism, which—as a general rule—favored natural over symbolical senses of scripture, encouraged natural historians of the sixteenth century to expunge the emblematic and hieroglyphic from their endeavors and to see creatures no longer as moral signs but as species open to inductive scrutiny. Instead of describing animals in terms of, say, their associations with classical gods, their appearance on ancient coins, the proverbs they excited, and the recipes they stimulated as well as their physical characteristics, students of natural history now progressively dispensed with such "sympathies" and "correspondences." As the English Reformers stripped icons out of sanctuaries and stripped allegory out of scripture, they likewise stripped symbolism out of nature. Thereby they helped lay the foundations for what we think of as modern scientific inquiry.

While the dominant streams in English science saw natural philosophy as a means of supporting Protestant Christianity, religion was not the only way local conditions imprinted themselves on England's scientific culture. Other regional peculiarities also made their presence felt. Not every social group in seventeenth-century England was assumed to value truth; truth telling was expected to display an uneven social distribution. It was the gentleman who constituted the culture's paradigm of the truth teller. The geography of credibility followed social contours. Because gentlemen enjoyed financial independence, they had no need to fabricate falsehoods. Save where there were very particular reasons for doubt, their word was their bond and to be taken at face value. Not so with other groups. Their economic subservience rendered the poor suspect as truth tellers. Merchants and traders were in the same boat: because their economic survival required material advantage, their word was not to be trusted. It was not that such groups routinely misinformed inquirers on matters of natural phenomena, of course. But being disbelieved carried far less social cost for lower social classes than for gentlemen. Anyone from this spectrum of society hoping to attain authority in matters of natural philosophy would find that adopting the civil conventions of the gentility helped enormously.

The codes of gentlemanly conduct and the genteel resources for warranting knowledge that governed English society at the time were thus important features of scientific practice. And it was out of these that Boyle and his associates carved the identity of the Christian virtuoso. This was the conscious cultivation of a social position that stood above the vulgar interests of corporations and institutions and consequently provided freedom to engage in the gentlemanly pursuit of truth. What that gentlemanly conduct entailed was spelled out in English courtesy texts of the day—books of etiquette educating readers in what was considered appropriate behavior in polite society. They required the genteel not to be too dictatorial, too extravagant, or too pushy in their claims to knowledge. Decorum allowed demurring without discourtesy. And this provided resources for a style of inquiry that promoted sober scientific exchange. Flashiness and extravagance were vulgar; sobriety and restraint were seemly. Intellectual good manners, it might be said, prevented precisely the kind of melee into which Galileo was plunged in Italy. To the extent that Italian science was a spectacular courtly affair, its English counterpart was a subdued gentlemanly pursuit.

It is clear, then, that distinctive scientific cultures developed in Italy, Iberia, and England during the sixteenth and seventeenth centuries. In these places there were crucial differences between what was investigated by students of nature, who had the standing to make knowledge, and what interests scientific projects were intended to advance. Circumstances differed elsewhere. In France the particular expression of the Counter-Reformation church, which did not quash new experimental endeavors, was an important feature, as was the network at whose hub sat the mathematician Marin Mersenne, who did all in his power to resist the revival of the idea that matter was infused with life forces. In Sweden, the peculiar alliance between the Lutheran Church and Aristotelianism, and the utilitarian mercantilism of the ruling Hat party, made for a distinctively patriotic science that gave pride of place to land survey, economic science, applied natural history, and investigations likely to deliver industrial innovation. In other regions, other conditions prevailed. All this means that "the history of the scientific revolution," conceived as a singular moment in Western intellectual consciousness, needs to yield to "historical geographies of scientific endeavors" in different regional situations.

Power, Politics, and Provincial Science

While it makes sense to speak of English science in the Restoration period or Iberian science under Manuel I, or Italian science at the court of the Medici, it would be a mistake to think that such descriptors imply entire stylistic uniformity or conceptual coherence. In different towns and cities, in different counties and provinces, in different municipalities and parishes, scientific endeavors have been molded by subregional particularities. By the same token, scientific pursuits have also been enlisted by certain groups as resources in campaigns of various stripes—to combat public unrest, to push for social reform, to counteract political discord. Either way, scientific subcultures have taken shape in response to the dictates of urban politics and industrial pollution or the demands of civic pride and radical protest. To get a sense of how provincial science may be shaped by the forces of political and social geography, I want to turn to circumstances in Victorian Britain. Several cross-cutting geographies are discernible here. On the one hand, different kinds of scientific endeavor were cultivated in different cities in response to the imperatives of local culture. On the other, certain sections of British society attempted to iron out the map of English politics—to outmaneuver social geography, as it were—by mobilizing "science" to moderate extremist radicalism of one kind or another. At the same time, the comfortable image of science as the pursuit of the gentlemanly classes was contested by seditious elements in society who, in their own spaces, cultivated a kind of science very different from that of the Tory establishment.

The making of "Manchester science" in the early Victorian period was inextricably bound up with municipal politics. In a city whose population increased fifteenfold in the half century to 1830, it was the energies of the merchant and manufacturing classes that stimulated economic growth. This new commercial elite, hitherto marginalized in the social order and keen to advance moderate political reforms, saw in scientific engagement the means of promoting "the democracy of the intellect." Because science could be used to support social progress and sustain an ethic of hard work, it became a major vehicle of cultural expression among those who wanted to counter Manchester's social isolation from metropolitan trends. In the hands of men like Joseph Priestley—chemist, radical theologian, and English Unitarian—it even had millenarian possibilities. For him the "social millennium" would be ushered in by the conjoint influences of commerce, Christianity, and "true philosophy." To Priestley, "the empire of reason" was nothing less than "the reign of peace." Science in this setting thus channeled the cultural values of the new local elites who patronized its institutions, attended its presentations, and in some cases participated in its investigations.

The shape of Manchester science, then, mirrored the changing topography of civic politics in Victorian England's second city. As such it bore the imprint of its industrial culture. In the early decades of the century, the atmosphere in Manchester's various scientific institutions, particularly the Literary and Philosophical Society founded in 1781, was that of the dilettantish gentleman-amateur. By midcentury, however, this style had given way to an unremitting scientific utilitarianism suited to the needs of the new middle class. High on the city's civic agenda were matters of public health and environmental quality in the wake of industrialization. Accordingly, the city spearheaded early research on atmospheric pollution, sewage disposal, general sanitation, urban overcrowding, and water contamination. Of central significance here was the work of statisticians and especially chemists, the most innovative of whom had studied in Giessen with the German chemist Justus von Liebig before taking up residence in Manchester in the 1840s. All of them dedicated their expertise in organic chemistry to the service of public health. Manchester science in this guise thus emerged as a species of civic virtue, and it became a strategic resource in moves for government reform. The sanitary chemist Robert Angus Smith, for instance, helped to bring problems of air pollution to the attention of government and to shape the findings of a variety of metropolitan commissions and public inquiries into such matters as noxious vapors and urban health during the middle third of the century.

In circumstances like these, political conditions directly impressed themselves on the culture of provincial science. Which subjects were chosen for interrogation, the social prestige attached to such undertakings, and the uses knowledge-making enterprises were put to all reveal that the making of Manchester science was political through and through. Elsewhere provincial science was also a resource for a variety of local cultural projects. During the first half of the nineteenth century, a period of local economic decline, that the Bristol Institution's conscious cultivation of forms of science that had no immediate utility was an expression of ebullient self-confidence on the part of the city's elite. By contrast, the Geological and Polytechnic Society of the West Riding intentionally departed from "polite" science in favor of practical applicability. At the same time, the Edinburgh Philosophical Society, founded in 1832, reflected bourgeois tastes and demands, while Newcastle's major scientific institution was intimately connected with the social networks of that city's dissenting substructure. The medical community in early nineteenth-century Sheffield was dominated by practitioners with radical or reformist political outlooks and Quaker or Unitarian sympathies in matters of religion—that is to say, they were "marginal men," still to achieve the social and professional standing they sought. Victorian Britain, then, displayed a distinctive "cultural geography of science." Bristol science, Manchester science, and Newcastle science are not the same as science in Bristol, science in Manchester, or science in Newcastle. The place-name adjectives in these designations attest to scientific practices that were constituted in different ways by different urban cultures.

Provincial science, then, followed the contours of Victorian Britain's political geography. At the same time, there were those who sought to enlist science in campaigns to even out the political landscape by defusing civic unrest and countering religious extremism. In their hands, science was promoted as a means of curbing seditious tendencies toward speculative politics and of supplying social cohesion in the face of both a rabble-rousing proletariat *and* conventional English social hierarchy.

Chief among these was the establishment in 1831 of the British Association for the Advancement of Science (BAAS). In an era of civil unrest in the wake of the Industrial Revolution, when workers were turning variously to Chartism or Methodism and provincial capitalists routinely opted for the new progressive science, the BAAS inaugurated its circuit of industrial cities. Aided by the communications revolution of the day—turnpikes, canals, passenger railways, and so on—the BAAS's inner core, the "gentlemen of science," found it possible to hold up a unifying moral vision under the banner of scientific neutrality. By representing a moderate and measured attitude toward political engagement, they believed they could use science to damp down the agitation generated by the new industrial order. Appealing to inexorable natural laws, these scientific champions were convinced that they could offer society a neutral means of communication that outflanked bigotry, passion, and sectarian zealotry. Party differences could be laid aside in the common search for the laws of nature. And by moving from city to city throughout the empire, the association could circumvent spaces of political resistance by nurturing "geographical union." At the BAAS the rising middle classes, the aristocracy, and the gentry could meet in congenial union to pursue universal scientific truth (fig. 22). In this way British Association science stood for all that was temperate, reasonable, and moderate. So even as it sought to erase the geography of political difference, the BAAS as a scientific consistory gave voice to its own geopolitics of science. It was a mobile space, seeking through its roving round to extend the dominion of religious liberalism, gentlemanly restraint, and social integration.

The BAAS's successes, however remarkable, were nevertheless anything but universal. For even as the organization was getting under way in the 1830s, a different politics was dramatically manifesting itself among certain medical practitioners as they encountered the pre-Darwinian evolutionary theories emanating from across the English Channel in Paris. British responses to these dangerously materialist speculations disclosed a distinct social geography too. Arriving first in Edinburgh, the new doctrine derived from the French naturalist Jean-Baptiste de Lamarck persistently turned up among one particular substratum of society, namely, working-class atheists who had found their way into marginal medical practices. Pre-Darwinian evolution thus was irresistible to those who scorned aristocratic privilege, and they deployed its scientific propositions about progress from below in the cause of political agitation.

When these evolutionary conjectures filtered down from Edinburgh to London, it is not surprising that they spread like wildfire among those young doctors who found themselves on the fringes

DINNER

GIVEN BY THE

Magistrates & Town Council of Glasgow,

TO THE

BRITISH ASSOCIATION FOR THE ADVANCEMENT OF SCIENCE.

—o—

LIST OF TOASTS.

No.	Toast	Proposed by
1.	THE QUEEN, ...	CHAIR.
2.	Prince Albert, ...	Do.
3.	The Queen-Dowager and the rest of the Royal Family,	Do.
4.	The Army and Navy, ...	Do.
5.	The British Association & the Marquis of Breadalbane,	Do.
6.	The City of Glasgow, and the Lord Provost and Magistrates, ...	MARQUIS OF BREADALBANE
7.	The University of Glasgow, and Principal M'Farlan, .	LORD BELHAVEN.
8.	The Scientific Institutions and Societies of Europe and America, ...	PRINCIPAL M'FARLAN.
9.	The Memory of James Watt, and the other eminent men of Great Britain who have contributed to the Advancement of Science, ...	GENERAL TSCHEFFKINE.
10.	The Noblemen and Gentlemen from other parts of the Kingdom, who have honoured this Meeting of the British Association with their presence, . .	SIR JOHN ROBISON.
11.	The Astronomers of the Continent, and Mr. Encke, .	PROFESSOR AIREY.
12.	The Royal Society, and its Noble Chairman, the Marquis of Northampton, ...	MR. ENCKE.
13.	The Foreigners who have contributed so much to the interest and success of this Meeting of the British Association, ...	MARQUIS OF NORTHAMPTON
14.	The Ladies who have honoured the Meeting by attending its Sections, ...	DR. BUCKLAND.
15.	Railway Communication, & other Improvements which tend to facilitate intercourse between mankind, and thereby promote friendly relations, . . .	LORD SANDON.
16.	The Members for the City, ...	CROUPIER.
17.	The Rajah of Travancore, the great Promoter of Science in the East, ...	SIR D. BREWSTER.
18.	The Commercial and Manufacturing interests of the Country, which owe so much to Science for their advancement, ...	LORD MONTEAGLE.
19.	Foreign Naturalists, and M. Agassiz, . . .	MR. LYELL.
20.	The Lord Lieutenant of the County, . . .	
21.	The Secretaries of the British Association, . .	
22.	The Local Officers for the Glasgow Meeting of the Association, ...	MR. PHILLIPS.

HEDDERWICK AND SON, PRINTERS TO HER MAJESTY.

22. A list of toasts by the magistrates and town council of Glasgow at a dinner on 23 September 1840 to honor the visit of the British Association for the Advancement of Science. The list illustrates the spectrum of society that the association's "gentlemen of science" wanted to unite under the banner of science.

of the medical establishment and among outcasts from the old-fashioned gentlemanly science of the day. Transformist, law-bound, deterministic construals of evolutionary processes could easily be enlisted to support radical assaults on professional injustice, political expediency, and a hierarchical social order bolstered by priestcraft, providence, and physicotheology. In this underworld of lowlife medicine, serviced by secular anatomy schools and radical nonconformist colleges, evolutionary theories easily gained a foothold. They became a means of challenging the Anglican Tory stronghold of the Royal College of Physicians and the Royal College of Surgeons. Thus Thomas Wakley's medical journal the *Lancet,* founded in 1823, denounced these establishment institutions as bigoted and irresponsible oligarchies in collusion with a cozy ecclesiastical establishment. Drawing support from dissident private teachers and radical general practitioners, the journal promoted both social and scientific secularism by flouting natural theology and championing mechanistic accounts of human anatomy along French lines. Wakley himself was hounded through the courts by hospital consultants claiming libel and piracy. As a radical member of Parliament for Finsbury, he variously attacked providence, the poor laws, and even the BAAS, describing its leaders as a swarm of spider-hunting wasps sucking dry the truly hardworking students of nature (fig. 23). In this way the language of evolutionary anatomy was exploited by its advocates to mount a powerful political assault on the medical elite of the day. Small wonder the altogether respectable Charles Darwin dithered for decades before laying *his* version of evolutionary theory before the English public.

In different regional environments, scientific enterprises disclose conspicuously different cultural politics. Cognitive styles have differed from place to place, as have the projects to which scientific practitioners devoted their energies. Similarly, science has served dramatically different agendas in different ideological spaces. Treating scientific knowledge as a universal phenomenon, untouched by the particularities of location, plainly will not do if we want to come to grips with the immense power it exercises in society.

23. A cartoon from an 1841 issue of Punch, *depicting Thomas Wakley, member of Parliament for Finsbury, as a jackdaw plucking fine feathers from the "Tory peacocks" on matters like the poor laws. Wakley's medical journal, the* Lancet, *was the chief organ in the campaign to undermine the Anglican-Tory medical establishment of the day.*

Region, Reading, and the Geographies of Reception

So far our concern with the cultural shape that science has assumed in different regional settings has focused on the production end of things. The consumption of science—the ways scientific theories and practices have been received in different arenas—also bears the marks of local circumstances. The same is true of scientific practitioners. Scholars have often had to turn nomad to escape local censure.

And they have not always been received in the same way. The degree of ecclesiastical surveillance, scientific patronage, and protective shelter has varied from region to region. The Spanish Inquisition, for example, made life difficult for anyone harboring ideas that seemed to endanger Catholic orthodoxy. In Poland and Hungary conditions were less severe, while in Germany religious fragmentation disabled any inclination toward a centralized system of censorship. In Sweden, Queen Christina protected freethinkers who sought refuge there, while Holland welcomed persecuted Protestants from France and Jews from Iberia. All this goes to confirm that Europe displayed a distinctive "intellectual geography" in the seventeenth century.

Like people, scientific ideas do not diffuse over a flat cultural plain. Rather, they are encountered in particular places. The meaning of particular scientific texts and theories has varied from place to place, and one way of uncovering such geographies of reception is to determine how various cultures judged certain works of scientific scholarship. Take the ways the writings of the Prussian polymath Alexander von Humboldt were received in different national settings during the first half of the nineteenth century (fig. 24). Significantly, the works from Humboldt's pen that received most attention in *his* day are not those that have been most visible to the eye of scholarship in ours. His major scientific conspectus, *Kosmos,* for example, attracted much less critical attention than his work on Mexico—the *Essai politique sur le royaume de la Nouvelle-Espagne* (1808–11)—perhaps because of the latter's commercial and geopolitical implications. Humboldt's rise to international scientific stardom, then, can be ascribed neither to *Kosmos* nor even to the highly significant three-volume *Relation historique du voyage aux régions équinoxiales du nouveau continent* (1814–31), but to his early contributions as a colonial surveyor.

This realization forces us to reconsider the meaning of Humboldt's significance in his own era. But even at the time, the way Humboldt was interpreted was far from uniform. In different places his work was construed very differently. English-language reviews of his Mexican writings were decidedly more critical than their French

24. An 1806 painting by F. G. Weitsch of Alexander von Humboldt in Venezuela. A diplomat, scientific traveler, experimenter, and man of letters, Humboldt has often been depicted as the last "universal scholar." Recent research on how his works were read in different national contexts, however, shows that Humboldt was construed very differently in different places.

or German counterparts and were much more prone to judge the work at the bar of natural theology. And then, whereas French and German periodicals tended to stress Humboldt's cartographic and geodetic contributions, the British were far more likely to reflect on the *Essai*'s mercantile and geostrategic implications for dealings with Asia and the Pacific Northwest. In these ways disparities between reviewing cultures did much to condition how their reading publics first encountered what historians have come to call Humboldtian science.

Textual significance evidently shifts from place to place and at a variety of scales. Distinctive cultures of reading can be detected within regions and between them, within cities and between them, within neighborhoods and between them. Something of the dynamic of these "geographies of reading" and their relevance for the reception of scientific ideas can be glimpsed by considering how, in different spaces, the controversial *Vestiges of the Natural History of Creation,* which first appeared in 1844, was encountered. An anonymous and controversial text, later acknowledged as the work of the Scottish publisher Robert Chambers, it caused a sensation at the time. The ways this pre-Darwinian evolutionary epic—which advanced a speculative developmental account of everything from the solar system to the human species—was read in different domestic, urban, and national settings tellingly discloses both the instability of textual meaning and a distinctive geography of textual interpretation. In different London salons and reading rooms, the book entered fashionable conversation in different ways and found itself very differently treated. Among aristocratic readers, such as in the home of Lord Francis Egerton—a leading Tory—it was regarded as poisonous, and the refutations streaming from the pens of scientific critics were warmly embraced. To the progressive Whigs who gathered in the drawing room of Sir John Hobhouse, it was boldly visionary and gloriously free of bigotry or prejudice. In Unitarian conversation, like that in the townhouse of Lord and Lady Lovelace on Saint James's Square, the book's emphasis on change from below was seen as a telling blow against a smug ecclesiastical establishment. Wherever polite conversation took place, *Vestiges* was talked about. After its publication, topics like human origins, which previously had been discussed only after the ladies left the men to their port and cigars, could now be brought up in mixed company.

Outside London, the book also enjoyed differing fortunes. Whereas in Oxford it was read as supportive of new scientific insights, in Cambridge it was vilified by writers like the clergyman-geologist Adam Sedgwick, who thought it an example of the most degrading species of materialism. In Liverpool, where it stirred up more sustained print controversy than anywhere else in Britain, the ways it was read mirrored the social microgeography of the city. It sold briskly among those pressing for urban reform, for example, because it could be interpreted as providing a scientific justification for social improvement. In Europe, *Vestiges* also enjoyed a wide readership in various translations. And here too the book's interpretative instability surfaces. The German version translated by Adolf Friedrich Seubert, for instance, strangely incorporated material—from William Whewell's *Indications of the Creator* (1845)—originally intended to rebut *Vestiges*. But by interlacing the two texts, Seubert succeeded in making *Vestiges* into a treatise confirming that evolutionary development took place according to divinely ordained laws. All in all, very different messages were read in, and read into, *Vestiges* depending on local circumstances. Textual meanings are mobile: they both create and are created by their own "geographies of reading."

Factors of this sort reemphasize the salience of regional traits in responses to scientific claims. To appreciate the power of place in sculpting encounters with science, it is worth pausing to ascertain how intellectual elites in different Victorian cities met the challenges of Darwinian biology. Even among groups with very similar religious convictions, it is possible to unpack local factors that fostered or frustrated the dissemination of evolution theory. By attending to these particulars, the differences that *made a difference* to the Darwinian diffusion begin to be exposed, as can be seen from the following "tale of three cities."

In 1874 church leaders in the Calvinist citadels of Edinburgh, Belfast, and Princeton pronounced on the new biology. In his inaugural lecture that October, Robert Rainy, the principal of the Free Church College in Edinburgh, openly accepted the legitimacy of evolutionary speculations—including the possibility that the human race had descended from animal forebears. At more or less the same time, across the Irish Sea in Belfast, J. L. Porter was telling his Presbyterian students that evolutionary theory threatened to quench every trace of virtue and that there was not a shred of evidence from which the pernicious dogmas of Thomas Henry Huxley and John Tyndall could be deduced. Just a few months earlier, on the other side of the Atlantic in Princeton, New Jersey, the doyen of American Presbyterians, Charles Hodge, urged in *What Is Darwinism?* that the rejection of divine design was the lynchpin of that system. And that was decisive. For, he insisted, it was the elimination of purpose and plan—not descent with modification, not species transmutation, not even natural selection—that brought Darwinism into conflict with Christian theism. All of this meant that it was entirely possible, Hodge reckoned, to be a Christian *evolutionist;* the idea of a Christian *Darwinian,* by contrast, was simply incoherent. To Hodge, Darwinism *was* atheism.

These key pronouncements were broadcast in differing ideological contexts. In each place, different issues were central in conditioning the rhetorical stances that commentators adopted in their evaluation of evolutionary theory. Besides, different voices were being sounded in different ways, and their modes of expression, whether bellicose or irenic, did much to set the tone of the local science-religion "encounter." In Edinburgh, evolutionary theories were rapidly domesticated to the needs of the Presbyterian establishment. This was very largely because the Darwinian issue paled in significance beside other intellectual currents assaulting the Scottish religious mind, most conspicuous among them the new biblical criticism that was beginning to receive an airing. William Robertson Smith, a professor in the Free Church College in Aberdeen, had revealed his support for these new currents in an infamous entry on the Bible for the *Encyclopaedia Britannica,* in which he allowed that the biblical text incorporated various ethnographic and mythological legends, and in a sequence of articles expounding the polyandric origins of Semitic marriage (fig. 25). His stance sent shivers down conservative Presbyterian spines, and he was dismissed from his professorship. With matters of this sort thrust onto their agenda, evolution posed little threat to a culture long enamored of scientific endeavors. In the years that followed a host of Scottish theologians gave their support to evolution in one form or another.

What contributed materially to the different ethos in Belfast was the coming of the Parliament of Science—the British Association for the Advancement of Science—to the city in August 1874. The president-elect was John Tyndall, and he took the opportunity to mount an assault on the old clerical guardians of scripture and social status in the name of the new priesthood of science. All religious theories, Tyndall proclaimed, must give way to the control of science. The gauntlet had been thrown down (fig. 26). Events moved quickly. Tyndall's address was the subject of a truculent attack by Rev. Robert Watts, who was already spitting blood since the Biology Section had turned down a paper he had prepared titled "An Irenicum, or A Plea for Peace and Co-operation between Science and Theology." It caused a local stir. Tyndall himself later reflected that every pulpit in Belfast had thundered at him. The BAAS event thus set the tone of the Belfast response to evolution for more than a generation. Nor was this just a short-term knee-jerk reaction. Twenty years later Watts was still reliving the events of that week in 1874; he neither would nor could release his grip on that bitter memory.

Yet for all their anxieties, Protestant critics in Belfast were unwilling to join forces with Catholic opponents of evolution. However similar in judgment and fearful of materialism they may have been, Protestants and Catholics alike used the Tyndall furor as an occasion to continue firing broadsides at one another. The Catholic hierarchy put the blame on the laxity of Protestant education and seized the opportunity to rebuke those who had become indifferent to the struggle for a Catholic system of education. For their part, Protestants cast secularization and Catholicism as subversive allies against the inductive truths of science and the revealed truths of scripture. They conflated, as a single object of opprobrium, an old enemy—popery—and a new

25. *Caricature of William Robertson Smith during his trial before the Free Church of Scotland. Here he is portrayed carrying a volume of the* Encyclopaedia Britannica, *where his controversial entry on the Bible was published in 1875. Evolutionary theory seemed relatively insignificant compared with the challenge that Smith's biblical criticism and anthropological speculations brought to the Free Church.*

26. Vanity Fair *cartoon of John Tyndall, president of the British Association for the Advancement of Science, whose "Belfast Address" in 1874 called forth condemnation from both Protestant and Catholic religious leaders in Ireland. Tyndall's attack made it extremely difficult for Ulster clergy to respond to evolutionary theory as positively as their colleagues elsewhere.*

one—evolution. Tyndall's speech succeeded not only in inciting the opposition of both Protestants and Catholics in Ireland to his own science but in furthering their antagonism to each other.

In transatlantic Princeton, things were different yet again. For alongside Hodge stood James McCosh, the new college president, who was determined to read evolution as the story of divine design. By keeping the rhetorical space of Princeton open to that possibility, he did much to determine that evolutionary theory would be tolerated in American Presbyterianism's intellectual heartland. Thus, over the following decades, a long succession of Princeton theologians stressed that evolution, at least in its more circumscribed form, was not incompatible with Christianity. Most conspicuous among these was B. B. Warfield, who, though providing an architectonic defense of the idea that the Bible was without error, went so far as to describe himself as a Darwinian of the purest water. The Princeton succession, retaining the cultural hegemony that neither their Irish nor their Scottish counterparts enjoyed in the years around 1900, benefited from an ecclesiastical control that enabled them to respond with equanimity to the Darwinian currents sweeping across the conceptual landscape.

In the latter decades of the nineteenth century, the intellectual leadership of the Presbyterian citadels of Edinburgh, Belfast, and Princeton were involved in the production and reproduction of cultural space. These maneuvers played a crucial role in the ways evolutionary science was encountered in these regions. In all three, the fields of discourse that religious leaders had done so much to manage set limits on the assertible—on what could be *said* about evolution, and on what could be *heard*.

Religious machinations, of course, were not the only conditioning factors in the regional rendezvous with evolutionary biology. In the American South, the antievolution sentiments of the Charleston circle of naturalists owed a good deal to southern racial ideology. The monogenetic implications of Darwin's understanding of human origins did not sit comfortably with the idea that the human race was composed of entirely different species, each of separate origin. Moreover, the enthusiasm of many southern naturalists for Darwin's most outspoken critic in America, the Swiss savant Louis Agassiz, who argued for a range of racial centers of creation, had an important influence. This is not to say that monogenists were never implicated in racial politics. In the case of the Charleston clergyman-naturalist John Bachman, a staunch adherence to the biblical unity of the human race did nothing to dilute his belief in racial hierarchy. But the willingness of the Charleston scientists to use natural history for racial purposes discloses the relevance of regional politics to the encounter with Darwinian theory. It was precisely because the racial obsessions of the Old South had secured the antebellum benediction of science that Darwin's account could now seem so threatening. Where southern opposition to Darwin did most forcefully surface was in matters to do with *human* origins. When Alexander Winchell lost his position at the University of Vanderbilt in 1878 over his suggestion that Adam had been preceded by preadamite humans, it was the implication that those forebears might have been black that contributed most to the furor. If that was where evolution led, the South definitely did not want to follow.

In New Zealand, by contrast, racial politics tended in a different direction. There Darwinism was espoused because it was seen as justifying an ethnic struggle for life and as legitimizing the settlers' routing of the Maori. Moreover, because religious ardor rarely rose above the lukewarm, New Zealanders responded with remarkable enthusiasm to Darwinism. The response of Canadians, in a context similarly concerned with assembling an academic infrastructure, was rather slower. Here the dogged digging for data—so strenuously underwritten by a flourishing Baconianism of Scottish derivation—together with Protestant-Catholic politico-religious struggles, meant that little time was left for theorizing of the Darwinian or any other variety. Besides this, the harsh physical environment of the Canadian North remained what one writer called "the single greatest fact" in the Canadian psyche. Endlessly resistant to agricultural taming and a monumental obstacle to northward settlement, it did a good deal to dampen nationalistic optimism at precisely the time the *Origin of Species* made its appearance. In these conditions nature seemed anything but a creative developmental force.

The vast expanses of a harsh, sparsely populated environment influenced the reception of Darwinism in Russia in a rather different way. Here Darwin's metaphor of a struggle for existence was resisted by the leading members of the Russian scientific intelligentsia. The St. Petersburg Society of Naturalists embraced versions of evolution that minimized the role of competition; they remained deeply skeptical of the Malthusian elements in the Darwinian scheme. In part this reflected the country's political economy, largely composed of peasants and landowners and lacking a market-driven middle class. In a political climate favoring cooperation, advocates of evolution aimed critical commentary at Charles Darwin, Alfred Russel Wallace, and unnamed "European Darwinists." Politically, they preferred versions of the theory in which "mutual aid" dominated. But the physical environment also had a role to play. A meager population and extreme climatic severity did not fit at all well with Darwin's picture of teeming life-forms or Wallace's lush tropical vegetation. Organisms in the Russian North were not packed into tiny, tight ecological niches. For Russian evolutionists, the Darwinian struggle just did not square with the Siberian land and climate; it seemed a theory made in, and for, the tropics. In Russia, Darwinism could survive only *without* Malthus, for both ideological and environmental reasons.

The reception of Darwinism thus displayed an uneven regional geography. In some cases religious commitment was crucial. In others racial neuroses or political fixations controlled the diffusion of the Darwinian mind-set. In yet others the contingencies of local physical geography were directly relevant. Whatever the particulars, local circumstances were decisive in shaping how regional cultures encountered new theories. In the consumption of science, as in its production, a distinctive regionalism manifests itself.

Science, the State, and Regional Identity

So far our reflections have centered on how the character of scientific inquiry, and responses to it, have been touched by regional culture.

But it would be wrong to think that the relations between science and region have been all one-way. Scientific knowledge and practice not only have been shaped by regional factors, they have also been instrumental in fashioning regional identity. Applied astronomy, precision mapping, resource inventory, and geodetic survey are just a few of the scientific practices that states have mobilized for the purpose of defining the bounds of its territory and providing a register of its natural assets. Such activities at once impose rational order on the seeming chaos of nature, give governments a sense of territorial coherence, and supply servants of the state with geographical data essential for fixing taxes, stimulating economic growth, exploiting resources, and maintaining military defense. Scientific endeavor is both a cause and a consequence of geographical agency.

The complicity of science in the constitution of senses of regional selfhood is particularly plain in enterprises that have had national labels appended to them—national laboratories, national surveys, national academies of science, and the like. Given its role in the genesis of the very idea of the "nation," it is not surprising that it was in France that the "national laboratory" made an especially early appearance. In the aftermath of revolution, when eminent scientists were swept into the service of the state at war, these institutions were geared to the needs of the military. Again and again, national laboratories have given expression to a craving for national unity and afforded the state the opportunity to put its technical glories on display.

In national surveys too the active agency of science in constructing state identity and in visualizing national space dramatically surfaces. Perhaps the earliest of these was the cartographic inspection of France that Louis XIV commissioned in the late seventeenth century to aid recovery from domestic disquiet and the wars with Spain. In so doing he hoped to unify provincial diversity under a strong central government. Under the leadership of the Cassinis—a four-generation dynasty of astronomers—detailed topographical maps of the territory were completed, using the latest astronomical techniques (fig. 27). Earlier, the contours of French terrain had been known only in "literary mode," that is, through lists of place-names, travel narratives,

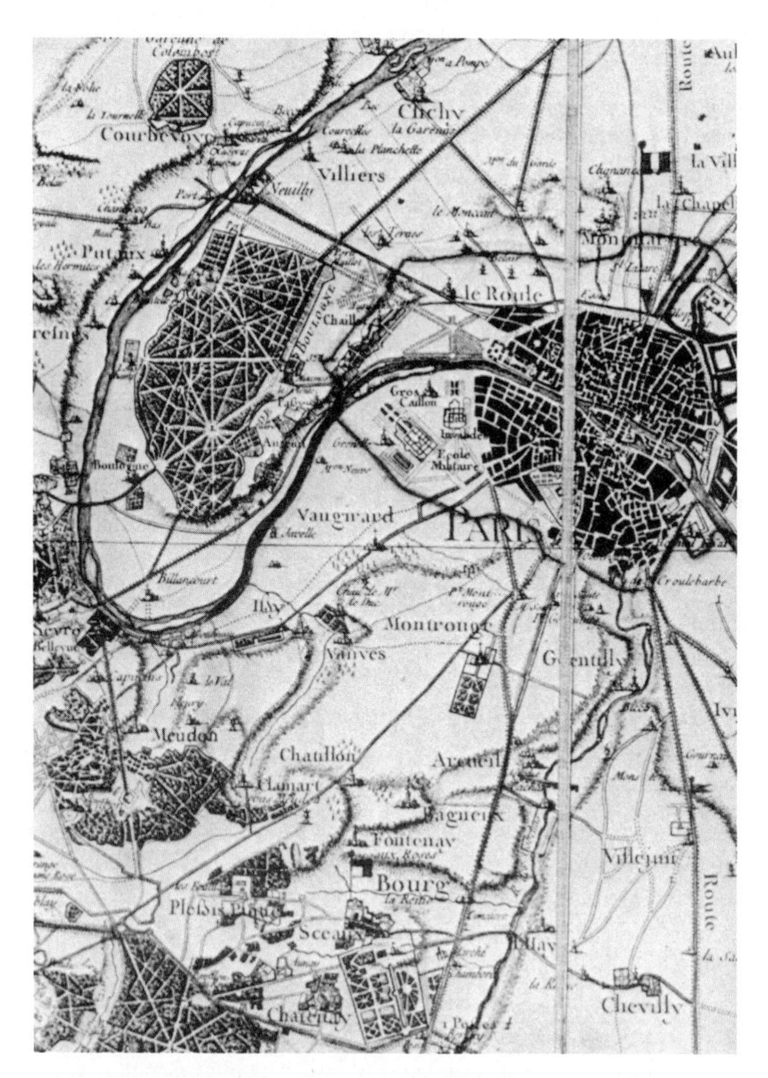

27. *A section from the topographical map of Paris and environs from the* Carte de Cassini, *published in 1793 and comprising 182 sheets at a scale of 1:86,400. The remarkable accuracy of the Cassini maps, which required astronomical precision and measurement standardization, were important components in the campaign to unify France under a strong central government.*

itineraries, and the like. Now scientific mapping provided a new means of collective spatial knowing suited to the needs of the state. Moreover, the local maps that did exist needed to be standardized and reassembled at the Royal Observatory in order to construct the country cartographically (fig. 28). By imposing national standards of measurement, scientific survey consolidated disordered space under the dominion of the monarch. In a sense, the map brought France into cultural circulation both on parchment and in perception. Not only did this achievement stimulate comparable efforts elsewhere, it also demonstrated the utility of the sciences of geography and cartography as handmaidens to state power. Survey lines on paper enabled the "rational" management of the nation's agricultural, economic, and natural resources. In Enlightenment France, then, science, survey, and a sense of nationhood were intimately interrelated. Concurrently in Scotland, geographical survey was regarded as so useful to the nation's sense of its self that in 1682 Robert Sibbald was appointed geographer royal. Sibbald's endeavors—the first of a sequence of such national geographical surveys—played a pivotal role in the making of Scottish national identity. By *recording* the established social order, scientific survey and cartography *reinforced* it.

Much the same was true of Jeffersonian America. For when Jefferson orchestrated the Lewis and Clark expedition up the Missouri in the early years of the nineteenth century, it was with the intention of bringing the American West within the tenure of the new nation's science. Like his own *Notes on the State of Virginia* (1780–81), the conception of this regional reconnaissance was patriotic to the core. With a loathing for Buffon's irritating allegations about the inferiority of the New World's environment and life-forms, Jefferson was determined to enlist science in the cause of loyal republicanism. Later surveys, like the United States Geological Survey, no less contributed to the American nation's sense of its continental identity. Indeed, what such surveys accomplished was a visualization of the state as a coherent geographical entity, imaginable, mappable, and therefore substantial. Through cartographic performance the very idea of a distinctive regional identity was rendered increasingly plausible.

28. The Paris Observatory, built between 1667 and 1672. Under its first director, Jean-Dominique Cassini, the first scientific survey of France was undertaken. When Louis XIV was shown the results in 1682, he was shocked to learn that the coastline of France had "shrunk" by more than a hundred miles in some places.

In these various national scientific enterprises, the power of scientific expertise to engender new forms of spatial consciousness and provoke new senses of geographical awareness is clearly discernible. What gives a state its identity, of course, is not just how it is visualized or constructed but also how it is regulated. Courtesy of the spirit of calculation and the impulse toward planning, the state has enlisted the methods of science not only in *making,* but also in *maintaining* national identity. So alongside its role in surveying the state's territorial scope and natural assets, scientific surveillance has been harnessed to manage cultural capital and demographic resources by applying quantitative procedures to public affairs. One or two moments will serve as indicators of science's complicity with what has been called "the scientific rationalization of society" and "governmentality," namely, the means by which everything from the self to the state has been subject to regulatory logic.

Consider conditions in seventeenth-century Germanic Europe in the aftermath of the Peace of Westphalia of 1648. In a region racked by religious strife, scientific knowledge and its technical applications were engaged as resources for reestablishing civic order and social discipline. In the face of political disarray, demographic devastation, economic recession, and a sense that society had lost its moral moorings, scientific principles were applied to solving everyday problems. Cameralism—as this impulse came to be known because the legislative council of regional rulers was often referred to as the camera—was geared to the rational organization and efficient management of the economy. It became the eighteenth-century Germanic science par excellence, with Joseph von Sonnenfels's *Principles of Police, Commerce and Finance* of 1765 establishing itself as the standard textbook for what was known as "the science of government." In this tradition, the means of studying the natural order were applied to the political realm, and cameralism thus culled insights from a mélange of such protodisciplines as agriculture, forestry, statistics, theoretical physics, and mining technology. Even when cameralist theories that harked back to past stability began to be eclipsed during the first half of the nineteenth century, the application of science to matters of state continued to flourish, not least in public health movements and urban planning enterprises that subjected society's everyday affairs to quantification. Thus Rudolph Virchow, a late nineteenth-century German professor of pathological anatomy who advocated both political and health reforms, insisted that scientific principles could be brought to bear on the rational upbringing of children. This regulatory thrust, of course, would soon find gruesome expression in various moves toward social hygiene and eugenic practices. In these and related ways, scientific ideology was enrolled in the service of state management and the reproduction of cultural identity.

In England this self-same inclination crystallized in what was known as political arithmetic, particularly as developed by the physician, land surveyor, and economist William Petty during the 1670s. Petty's aim was to deliver an exhaustive computation of England's demographic and capital assets. A decade and a half before he took up the task of writing *Political Arithmetick* (which was not published until after his death), he had been urging Charles II to compile a land registry incorporating population data. Not surprisingly, as a fellow of the Royal Society Petty drew on scientific modes of analysis as he pursued his task. Political arithmetic, as he conceived it, was to be nothing less than government enacted on Bacon's principles of scientific method. Not only did his enthusiasm for experimental mechanical philosophy manifest itself in his turn toward social quantification, it also convinced him that human activities, like the material world, were governed by inexorable natural laws. Besides, he routinely appealed to medical metaphors in his treatment of what he took to be society's ills. In consequence, as with scientific inquiry more generally, the import of Petty's social project was toward what we might call the demystification or disenchantment of the world. Rather than attributing the commercial success of other countries to some "national spirit" or what he dismissively called "angelic wits," he was inclined to turn to such mundane matters as geographical location, trade patterns, and shipping tonnage. It was a thoroughly empirical move that, like the demographic work of John Graunt in 1662, used aggregate data in the attempt to extract previously unobserved regularities hidden beneath chaotic and messy surface appearances. Fundamentally this involved "objectivizing" human beings and regarding them as commodities whose value could be expressed in monetary terms. In Petty's hands, the political arithmetic undertaking was intended to equip the sovereign state with scientific means to enhance the standing of the commonwealth and the affluence of its citizens. Its colonial potential was most forcibly made manifest in Ireland, where, while serving as physician general of Oliver Cromwell's army during the mid-1650s, he conducted the first large-scale scientific land survey. In the hands of figures like Petty, the computational methods of natural philosophy were enlisted in the service of state management and the organization of national space.

The cameralist disposition to bring scientific precepts to bear on matters of state did not invariably express itself in the language of quantification, however, and nowhere is this more clearly exposed than in the economic policies of the famous eighteenth-century Swedish botanist Carolus Linnaeus. Linnaeus's scientific reputation rests very largely on the taxonomic scheme he devised for classifying organisms by using a two-word designation comprising genus and species. This binomial system of naming flora and fauna, together with his sexual means of classifying plants, established his reputation as the greatest natural historian of the Enlightenment. But Linnaeus always conceived of himself as fundamentally an architect of the state, and his mind moved fluidly between what he called "the economy of nature," the divine economy, and national economic policy.

Linnaean-style cameralism needs to be seen in the context of Sweden's postimperial circumstances. In the course of the Great Northern Wars with Russia during the first two decades of the eighteenth century, Sweden had lost its Baltic colonies, and the country's elite determined that their economic future lay in internal development. In exchange for an extensive empire, they sought a sturdy nation, and they turned to science as the means of achieving this ambition. Indeed, the founders of the Swedish Academy of Science initially dubbed it the "Economic Society of Science." The patriotic

Linnaeus warmly embraced this vision. His strategy was to urge that Swedish autonomy could be secured through two related ecological policies. The first tactic was to carefully attend to the natural resources that the divine economist had distributed to the nation's own region and to impose tariffs and levies on imported goods. The second method was to seek to reassemble the world's plant riches within Sweden itself and then subject them to a careful program of acclimatization. Transplantation and naturalization were the keys to self-sufficiency and thus to national prosperity. Accordingly, on their voyages Linnaeus's disciples were charged with the task of garnering the earth's useful herbs and plants. Growing tea in Sweden, he insisted, was as momentous an achievement as winning a war. National development, he was sure, was therefore not about territorial acquisition; it was about ecological enrichment. Botanical science could literally remake the nation's biogeography. To Linnaeus, economics was simply the science of how to harvest nature.

Besides these more or less direct scientific interventions in state policy, the ideology of science has also been of importance in efforts to secure political cohesion and identity in a variety of situations. Throughout the seventeenth century, the social utility of science was widely felt in an era of religious enthusiasm and ecclesiastical fragmentation. Courtesy of the conviction that science, unlike politics, was seen as a guarantor of universal truth, the widespread practice of scientific inquiry was encouraged as a means of maintaining social order and moral authority. Thus in the hands of Bacon, England's chief apologist for "scientific method," the advancement of learning (to use the title of his 1605 treatise) was harnessed in the cause of a Protestant culture fearful of the sectarian disruption so typical of its Continental counterpart. Baconian inquiry was intended to preserve national stability through the conjoint energies of scientific gentlemen and mechanical artisans working together for the improvement of citizens. To Bacon, the beneficial scope of the new learning seemed limitless. It delivered ecclesiastical, technological, economic, and political goods to the nation.

As the century wore on, England found itself host to a medley of radical reformers and sectarians pushing for the redistribution of wealth, the licensing of women preachers, wider democratic representation, the reallocation of property, and the like. Many of them drew support from the idea of nature as spiritually animated and possessing inherent forces. In this environment, moderate reformers turned to the Newtonian mechanical philosophy as a means of curbing radical dissent. Why? In a world where natural philosophy, religious creed, and political authority were intimately interwoven, ideas about matter really mattered. Newton's universe was promoted as an alternative to two extremes. On the one side there was the philosophy of those like René Descartes, who conceived of matter as composed of nothing but an infinitude of tiny particles or corpuscles. That view just seemed to banish the Creator from his creation. While Descartes himself saw his system as supporting orthodox Christianity, English critics sensed in it a materialism that was next-door neighbor to atheism. On the other side were what Newton called "vulgar" notions of vitalism and pantheism. According to these matter was inhabited, in one way or another, by spiritual forces and occult powers—the kind of animated cosmos postulated by natural magicians. Nature thought of in this way could even be credited with sentience. Over against both materialism and mysticism, Newton's mechanical philosophy insisted that matter, though inert and understandable in the language of mechanism, all the while bore witness to the wisdom of God. By studying nature, natural philosophers were studying the handiwork of a rational Creator. Matter conceived of in this way could not be called on to support the pantheistic and revolutionary inclinations among some Ranters, Diggers, Levellers, and other red-hot Protestants attacking traditional ecclesiastical and political authorities. Still less could it justify the seditious politics of atheists and materialists that Newton himself abominated. On the contrary, Newton's universe, providentially directed, restored rational order to nature. Precisely the same principles should also govern church and polity. In this context different understandings of matter and of God's role in overseeing the natural world were part and parcel of political discourse on nation, state, and authority, not least because the way God regulated nature was analogous to the way a monarch ruled his kingdom. Natural philosophy Newton-style was thus a key weapon in the arsenal of those who wanted to combat the fragmentation of England's political geography and to impose rationality on its spaces of social disorder.

In other contexts, science was used in different ways to underwrite state ideology. In late nineteenth-century Argentina, for instance, the spread of scientific education was seen as crucial to national recovery. It could provide an inventory of natural resources, subvert the Scholasticism that still lingered in educational institutions, and place the nation's history within the optimistic framework of inexorable social improvement underwritten by evolutionary progressivism. Science, in Argentina, was espoused as the means of escaping economic backwardness and creating a modernist cultural identity. Again in the Soviet Union, the official communist adoption of the evolutionary thinking of the agronomist T. D. Lysenko in the 1930s is notable. His ardent advocacy of the idea that acquired characteristics could be inherited proved ideologically irresistible, especially when he claimed it could be successfully applied to remedy Russia's chronic wheat shortages in the wake of the catastrophic collectivization policies of the previous two decades. For apart from the idea's promised agricultural benefits, which never materialized, it resonated with Marxist hostility to the seemingly heartless capitalism of natural selection. Under Stalin's regime, Lysenko became director of the Soviet Academy's genetics institute, and from that position he banished many scientists who did not share his views. Here state creed and identity received the support of, and in turn conferred official approval on, a highly idiosyncratic version of evolutionary biology.

In a range of different ways, then, scientific practices have been enlisted in the service of the state. They have been implicated in the fashioning of national identities through the crafts of geographical survey. They have been entangled in the regulation of the state through various methods of social surveillance. They have been used as a resource in campaigns to undermine revolutionary elements in society. Science has thus been actively engaged in the shaping of regional cultures even as it has been shaped by them.

* * *

Science has been, and continues to be, promoted as a universal undertaking untouched by the vicissitudes of the local. Our travels in this chapter, however, have exposed something of the degree to which scientific endeavor has persistently exhibited distinctly regional features. Science has borne the stamp of the regional circumstances within which it has been practiced. At the same time, regional cultures have had a profound influence on the reception of new theories and on the rhetorical stances adopted by interlocutors in public debates over scientific judgments. All the while, the ideology and practices of science have frequently been deployed in efforts to fashion and fortify identity at state and provincial levels. If we are to make sense of those practices called "science" as a dominant feature of the culture of modernity, then we will have to take with much greater seriousness "the regional geographies of scientific endeavors."

Circulation

MOVEMENTS OF SCIENCE

On 4 May 1827 Étienne Geoffroy Sainte-Hilaire arrived in Marseilles to take delivery of a gift that Muhammad Ali, the Ottoman viceroy of Egypt, had presented to King Charles X of France. For several days he looked around the city's museums and collections and conversed with its leading savants. Now, early on the rainy morning of Sunday 20 May he departed with the Muslim ruler's present wrapped in oilskins and accompanied by an entourage that included two mouflons—wild mountain sheep. It must have been quite a sight. For Muhammad Ali had presented France with its first giraffe (fig. 29). Captured when she was scarcely two months old, she had been in transit from southeastern Sudan for two and a half years—including some three weeks on the Mediterranean, lodged in the hold of a brigantine with a hole cut in the deck through which her head protruded. Now, dressed in a black raincoat, she began the final leg of her journey, the 550-mile, forty-one-day walk to Paris. Such was the stir the whole mission caused that some 30,000 spectators trooped out to see her on her way through Lyons; during the following summer, well over three times that number visited her in the Paris Jardin du Roi.

29. A portrait, by Nicolas Huet, of the Sudanese servant Atir and the giraffe presented by the Ottoman viceroy of Egypt to the king of France. They walked from Marseilles to Paris during May and June 1827.

If the giraffe event was in some sense an exercise in international diplomacy by an Egyptian francophile who had long cultivated French connections, it was also a further chapter in Europe's scientific appropriation of the East. As French savants flocked to see this latest exhibit of Oriental exoticism, they also continued from afar the Napoleonic intellectual conquest of Egypt and confirmed the importance of trafficking in animals as they pursued scientific knowledge of the organic world.

More or less similar enterprises flourished elsewhere. In 1830 a couple of English surveying vessels had been in the South Atlantic for three or four years establishing accurate latitudes for South American sites and plotting the intricate coastline between Patagonia and what was known as "the land of fire"—Tierra del Fuego. After all, this was one of the world's most strategic shipping lanes, and through the medium of maps, the Royal Navy believed it could exercise dominion over the southern seas. But when HMS *Beagle* pulled away from Tierra del Fuego in late May or early June of 1830, it had more than new survey charts on board; four native Fuegians were also beginning their long voyage to the other side of the world. Originally taken captive in feuds with the local people, they became in Captain Robert Fitzroy's eyes an experiment in the powers of civilization. Believing that they would benefit from exposure to English habits and that their return would have a transforming effect on a savage Fuegian society, Fitzroy undertook to have them educated in English ways and then restored to their homeland in 1833 when, now accompanied by Charles Darwin, the *Beagle* again found itself in southern waters. It was a disaster. The Fuegians had rapidly and happily adjusted to the fashionable niceties of Victorian high culture. They just as easily reverted to their "savage" state within days of setting foot on their native soil. Two of them turned on the third and stripped him of everything he possessed, down to the kid gloves and button boots in which he had taken such foppish pride. And yet when the victim was offered the chance to return to England, he replied that he had no desire to do so. Civilization, it seemed, was a fragile thing and no match for the awesome power of social and physical environment. Only an experiment in geographical transplantation could have delivered such distressing findings.

The importance of circulation in the geography of science is not restricted to the movement of species and specimens, of course. Ideas and instruments, texts and theories, individuals and inventions—to name but a very few—all diffuse across the surface of the earth. Take the spread of the Copernican theory throughout Europe during the early seventeenth century. The locations of Copernicus's *De Revolutionibus* (in both the 1543 Nuremberg and 1566 Basel editions) by 1620 provide an initial clue to the diffusion of the heliocentric system. Moreover, because a papal decree was issued in March 1616 that specified a number of alterations to be made to the text to comply with Catholic orthodoxy, it is possible to identify where unexpurgated and censored versions of the treatise turned up. What immediately becomes clear is that while most Italian copies were censored, the Decree of the Holy Congregation had relatively little impact elsewhere. Even in France, where most copies were in Jesuit libraries, there is little evidence of censorship, perhaps because the Jesuits considered efforts at suppression a Dominican obsession. Of course the diffusion of Copernicanism cannot simply be "read off" censorship cartography. Other factors had important roles. The slow headway that the Copernican theory made in Scotland compared with England, where it gained an early foothold, had as much to do with the country's political unrest as with the relative absence of indigenous astronomical publication. In the Netherlands, the association of heliocentricity with certain brands of Protestantism was strong enough for one opponent to refer to it as the "Calvinistic-Copernican system." Whatever the details, the new astronomy diffused unevenly, with distinct spaces of resistance and support.

Comparable stories can be told about other components of scientific endeavor. Technical equipment, for example, is also mobile. Thus in the 1660s various efforts were made throughout Europe to construct replicas of Robert Boyle's celebrated air pump, which had been invented to produce a vacuum in its glass receiver by expelling air with a piston. The appliance was hugely significant, but not simply as a contested means of producing a vacuum. It was emblematic of the new experimental method and of the idea that nature could be known through human artifice. Duplicating the instrument, however, turned out to be far from easy, and problems of replication were manifold. We will turn to the significance of these difficulties presently, but for the moment I simply note that efforts to construct the new machine were made in Paris, The Hague, Würzburg, Florence, and several other sites. At the same time, the diffusion of the apparatus from center to center gave expression to a new philosophy—that matters of fact could be delivered through experimental means. This may seem obvious to us now, but in the seventeenth century the suggestion that *natural* facts could be *artificially* disclosed had to be fought for. The dissemination of Boylean equipment was thus the recreation in different locations of a new kind of interrogatory space in which the disciplined manipulation of nature could take place. As the apparatus circulated from place to place, it trailed with it a philosophy concerning how best to find out about the natural world.

The list of items of scientific circulation and the means of transmission could go on and on. Scientific societies, learned academies, field clubs, and circulating libraries, as cultural innovations, have spread from one place to another. This happened not least during the period of the European Enlightenment, when salons of polite discourse, a mushrooming print culture, and coffeehouse sociability became conspicuous features of the public sphere. The Royal Society of London, for example, received its charter in 1662, and the Académie Royale des Sciences came into being in Paris in 1666. Over the next century or so scores of similarly inspired institutions were created in places like Berlin, Philadelphia, Boston, Saint Petersburg, and Stockholm. And alongside these organizations, peripatetic mathematical practitioners, public lecturers, merchants, itinerant clergymen, journalists, and a host of others were conduits in the flow of intellectual capital. New technologies and their accompanying mechanical skills likewise moved from site to site. In these and dozens of other ways, scientific knowledge was dispersed. Providing a catalog of modes of scientific diffusion or of the capillary networks through which scientific knowledge coursed, however, is not my concern here. Instead I want to dwell on the *conceptual* significance of circulation for scientific inquiry in order to tease out just how profound the influence of geography has been in the production of scientific knowing and in its movement around the globe.

Translocation and Transference: The Problem Stated

Many of the key conceptual issues having to do with knowledge and circulation revolve around two connected points. These concern, first, the ways scientific knowledge moves from place to place and, second, the means by which knowledge gleaned in faraway places travels back home. How is it that science, given the local dimensions we have already explored, travels across the surface of the earth with such seemingly effortless efficiency? And how is it that we acquire knowledge of distant peoples, places, and processes when the eyes and minds and bodies of others—not ours—are necessarily involved in firsthand witnessing?

The success of science in moving from location to location makes it altogether remarkable. But just how *do* scientific propositions, perceptions, and procedures migrate from their place of origin to radically different environments and find ready acceptance there? The usual answers are that scientific knowledge is transcendent, neutral, and disembodied; that its claims have ubiquitous validity; and that its diffusion is simply a consequence of its inherent universality. When French scientists repeat experiments carried out in California, they get the same results because natural laws operate the same way in Stanford and Paris, because science has taught us how to correctly interrogate nature, and because the scientific community keeps a check on things to ensure that the proper procedures are employed. Indeed, we are told, this is exactly what distinguishes science from folklore, politics, poetry, faith, or mere ideology.

But is this true? Is the transmission of scientific knowledge such a straightforward thing? Take the duplication of experimental findings. The capacity to reproduce the results of experiments carried out elsewhere obviously requires the appropriate apparatus. But duplicating equipment was never a simple task. Consider Robert Boyle's air pump (to which I have already referred) and its replication in the 1660s. For a start, virtually all the air pumps constructed in the decade immediately following the advent of Boyle's machine required their makers to see his prototype firsthand. Boyle's written account was never sufficient for transmission of the device. Translocation was therefore no simple matter; it required the transfer of hands-on craft competence. Moreover, all air pumps in the period gave experimenters trouble in one way or another. They had to tinker with the size of the glass globe and with the valves, the pistons, and other design features. This meant that the air pump was in constant alteration: transmission meant transformation. In fact, determining that a machine was in good working order was an intensely troublesome task. What passed as evidence that the air pump was doing its job? How could mere anomaly be discriminated from a matter of fact? All these questions bore on the spread of the appliance, because circulation required calibration. And here disputes arose. To say that a machine was a good one only when it delivered the results Boyle had achieved was tricky, since the whole point was to put his results to the test. Not surprisingly, the findings that a rival like Christiaan Huygens thought confirmed the worth of his own air pump disqualified it in Boyle's eyes. The translocation of equipment plainly did not mean the transference of findings. The dissemination of facts was simply not reducible to the migration of instruments. Yet the new experimental philosophy that Boyle advocated could travel from place to place only when his experimental space (its gadgetry and its accredited company of observers) was reproduced in different locales. And that meant using Boyle's results to calibrate the very machines that were intended to assess the validity of his findings.

This particular case brings to the surface at least two vital considerations in thinking about the migration of science. First, the diffusion of mechanical contrivances is never sufficient to ensure the unproblematic replication of any particular scientific proposition. But second, even where findings are reproduced, it is not unreasonable to ask questions about the connections between the calibration of apparatus and the nature of discovery. The production of experimental facts is inescapably tied to the reproduction of equipment, with all the circularity inherent therein. In a fundamental sense, laboratory knowledge is local knowledge. It is bound up with particular practical know-how, with the on-site availability of appropriate bits of technology, and with knowing one's way around machines. Knowledge acquired in this setting depends on "craft knowledge" of the workings of experimental devices. And its circulation beyond the confines of one venue is not simply the story of universal truths being manifest in particular settings. It has also to do with managing the transfer from one local venue to another. The world of facts that is generated by equipment constitutes the proximate data that scientific claims refer to. Without instrumental reproduction in other locations, "findings" would not be found. So whether because of the difficulties of duplication or because replication is required in the first place to reproduce data, the geographical spread of experimental knowledge is a more complex suite of operations than might at first appear. What looks like the universalism of science—its seemingly problem-free transferability from one arena to another—turns out to have much to do with the replicating, standardizing, or customizing of local procedure. Scientific knowledge gleaned in laboratories is thus less about the local instantiation of universally valid facts than about what one writer calls "the adaptation of one local knowledge to create another."

The circulation of scientific knowing, of course, is of wider dimensions than the replication of laboratory instrumentation. Take sciences like observational astronomy, geography, natural history, surveying, meteorology, hydrography, and medicinal botany. Their development has been inextricably bound up with traveling to distant realms. Such exploits involved hundreds of people engaging in spatially and temporally extended projects. In sixteenth- and seventeenth-century Europe—at a time of hitherto unprecedented global mobility—the empirical riches and conceptual challenges that arose from geographical reconnaissance played a profound role in a variety of scientific ventures. Francis Bacon himself advertised this connection when he mused that it was through "the distant voyages and travels which have become frequent in our times [that] many things in nature have been laid open and discovered which may let in new light upon philosophy." Evidently, for Bacon the new developments in natural philosophy were intimately tied up with a new geographical sensibility.

The list of scientific pursuits benefiting from distant data is not a short one. Edmund Halley traveled to St. Helena in 1676 to observe a lunar eclipse. In France, Jean-Dominique Cassini correlated astronomical observations from a wide range of informants to produce his famous terrestrial planisphere originally depicted on the floor of the Paris Observatory (fig. 30). Robert Boyle had to rely on data from abroad to test his hypothesis that the specific gravity of certain natural objects was geographically variable. The tables on which Isaac Newton based his amended computations of the orbit of comets in the second edition the *Principia* came from observers in different hemispheres. John Ray's *History of Plants,* which came out between 1686 and 1704, drew on observations made by botanical travelers across four continents. Such activities proliferated. Experiments with the barometer and the pendulum were conducted on far-off mountaintops. Botanical specimens and other materials flooded back into European gardens and salons. So too did the images that illustrators produced. Asian knowledge about pharmaceutical and therapeutic subjects made its way into European medical thinking through the anthologies compiled by scientific travelers. Soon scientific voyaging would become such a well-established mode of inquiry that James Cook, Jean-François de La Pérouse, Alexander von Humboldt, Charles Darwin, and many more became household names. Through such expeditions a worldwide network of centers was established, providing data on everything from terrestrial magnetism to zoological species. In all these ways, scientific knowledge in Europe depended on global circulation, and domestic maps of knowledge were continually reoriented in the light of the faraway.

Yet reports from afar created as many problems as they solved,

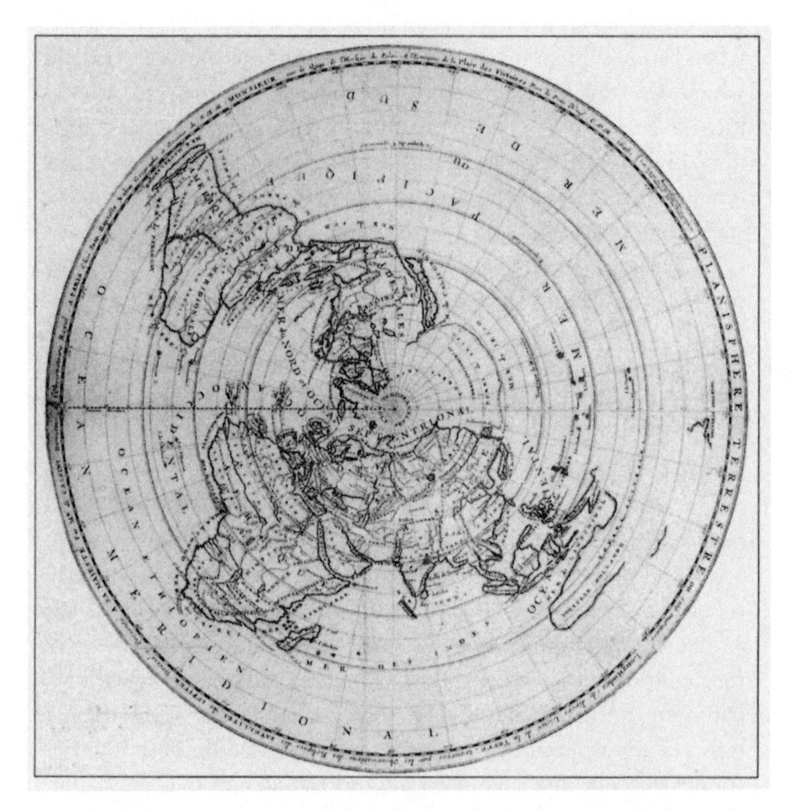

30. The Planisphere terrestre *by Jean-Dominique Cassini, originally depicted on the floor of the Paris Observatory. Producing this terrestrial image required compiling a wide range of celestial observations collected all over the world and collated at the Paris Observatory.*

and two in particular quickly surfaced. First, the disclosures of seafaring eyewitnesses profoundly challenged ancient authority. Latter-day travelers saw people and plants and places about which the ancients were in complete ignorance. No longer could Aristotle or Pliny or Ptolemy be unconditionally relied on. As some writers reflected during the early decades of the seventeenth century, global exploration had destroyed the foundations ancient philosophy had rested on, and a radically new conception of things was therefore inescapable. Moreover, travelers' tales compelled Europeans to compare the manners and mores, religions and regulations of different peoples. Suddenly, as Paul Hazard observed, "concepts which had occupied the lofty sphere of the transcendental were brought down to the level of things governed by circumstance. Practices deemed to be based on reason were found to be mere matters of custom."

Of just as great significance was a second challenge. Knowledge derived from travelers' experience necessarily created problems for the *ways of knowing* that the new champions of natural philosophy vigorously promoted. They had insisted on the all-importance of eyewitnessing, of direct experience, of immediate sense perception. William Harvey, who famously set forth his account of the circulation of the blood in 1628, urged his students not to rely on the experience of others and to take nothing on trust. In his *Sylva* of 1664, John Evelyn disparaged those works that rested on trust in other writers. Such proclamations were intended as a radical departure from earlier ways of knowing. In the *Confessions,* which he composed in 397–98, Augustine had recognized the inevitability of placing trust in other witnesses: "I began to realize," he wrote, "that I believed countless things which I had never seen or which had taken place when I was not there to see—so many events in the history of the world, so many facts about places and towns which I had never seen, and so much that I believed on the word of friends or doctors or various other people. Unless we took these things on trust, we should accomplish absolutely nothing in this life." Or again, the essayist Montaigne had insisted in the 1580s that "almost all the opinions we hold are taken on authority and trust." Now, in contrast, knowledge was to be placed on a surer foundation. It would be built on experience rather than authority, on witness rather than report, on observation rather than trust.

Despite their rhetoric, however, the new natural philosophers—again and again—could do no other than depend on the testimony of others. Boyle, for example, had to rely on the recorded witness of divers (using a diving bell) to test his ideas about the "weight of the air" even though he hedged the whole account about with various *ifs* and *buts*. Again he had to depend on the observations of travelers to polar realms to determine the influence of the cold on natural bodies. And yet he routinely insisted that the facts he delivered were restricted to those things to which he was an eyewitness or in which he himself was an actor. Similarly, the seventeenth-century Dutch natural philosopher Christiaan Huygens had to take on trust the reports from sea trials on the reliability of chronometers. When he couldn't bring himself to accept some particular result, he often put it down to the wearying effects of seasickness. And the nineteenth-century astronomer John Herschel urged that the only way that knowledge of terrestrial magnetism could be acquired was by collating observations made in every region of the globe. The need to exercise faith in others for some empirical findings was inescapable. Yet it was often taken as an unfortunate state of affairs in the kingdom of knowledge. And acquiring information about the natural history and geography of distant lands was even worse, for domestic knowledge of the remote relied almost exclusively on the testimony of others. In weighing testimony, of course, issues of judgment predominated. Who could be trusted? *That* was the question. And answering it was as much a matter of judging the integrity of people as of comprehending methodology or data. Finding out about distant *things* required discernment about *people*. Knowledge of nature and knowledge of people were joined at the hip simply because the processes of achieving warranted credibility have always been resolutely social.

There were other ironies too. Not least was the fact that the published accounts of scientific travelers were rarely composed with fresh brine on the brow. They were usually the product of lengthy compositional revision. James Cook, for example, repeatedly reworked his own manuscript narrative, thereby distancing it from the immediacy of the very circumstances it professed to disclose. Moreover, the final published version was the product of further editorial refinement by John Douglas, who drew on the rather different descriptions in the diaries of other officers on the ship. Seen in this light, any seeming experiential spontaneity was as much the outcome of editorial fashioning and rhetorical flourish as of direct empirical description. When we further reflect that documents like these were read by later travelers to prepare themselves for distant journeys—indeed, that they frequently accompanied scientific explorers to distant parts—the ways travel narratives were a composite product of stylistic convention, personal experience, and travelogue heritage becomes clear. Humboldt's *Narrative,* for example, was Darwin's constant companion on his five-year round-the-world voyage.

The circulation of scientific knowledge, then, raised profound cultural and conceptual challenges. So it is not surprising that persistent questions soon arose about just how to manage knowers-at-a-distance. How could the trust relationship inherent in the inescapably geographical character of scientific circuitry be made to bear the weight of the term "knowledge"? What mechanisms could be put in place to guarantee the reliability of those claims that floated in with the tide? The "techniques of trust" that were mobilized to minimize the risk inherent in listening to, and believing, voices from afar will appropriately be our next port of call.

Travel and the Techniques of Trust

It was notoriously hard to have complete confidence in reports from faraway places. Just how could one distinguish honest travelers from travel liars, faithful witnesses from fanciful storytellers? Essential to the circulation of scientific knowing, therefore, was the need to find means of overcoming such problems. How could knowledge merchants be governed in such a way that they would dependably act at a distance? How could both travel and travelers be *regulated* to ensure reliability? Various methods were put in place. And in each case the aim was to bridge the cognitive gap between presence and absence. Those *absent* from some space of knowledge production needed to find ways of assuring themselves that those *present* had gathered information in an appropriate manner. So we turn now to some of the techniques used to circumvent this species of difficulty.

DISCIPLINING THE SENSES

At the most basic level, the simplest way of guaranteeing the trustworthiness of knowledge collected far away is to ensure that observations are carried out by properly trained eyewitnesses. By disciplining the senses of observers, by supplying them with suitable instruments, and by instructing them in the techniques of data gathering, much of the space between "here" and "there" could be spanned. When the expense of specially commissioned overseas expeditions became too great a financial drain on the resources of the Académie Royale des Sciences during the late seventeenth century, Jesuit missionaries were trained and equipped in such subjects as astronomy and mathematics for the purpose of conducting cartographic surveys and related schemes. Armed no less with thermometers, air pumps, and instruction manuals than with the Bible and religious tradition, Father Guy de Tachard and six associates headed off for China in 1685. From their mobile laboratories, they sent back information on lunar eclipses, reports on the accuracy of longitude clocks, nautical data, botanical specimens, geographical digests, and much more besides.

Disciplinary techniques of this stripe had actually been in operation for quite some time and would certainly continue to be used for generations. A few moments—mostly from the sixteenth and nineteenth centuries—will illustrate something of the maneuvers involved in projects designed to transcend the inherently geographical problem of information circulation between home and away.

With the opening up of the world to European eyes, countries like Portugal were faced with the problems of sustaining an increasingly global, seaborne empire. While developments in ship-building technology and the like were undoubtedly crucial, it has become increasingly clear that managing people was just as critical. As seafarers strayed farther and farther beyond familiar waters, the navigational traditions they had hitherto relied on became progressively unreliable, and new techniques had to be brought into play. To be sure, scholars had supplied theoretical solutions to such problems for long enough. But providing ships' captains with astronomical information in an accessible and pragmatic form was a different matter. The Portuguese triumph over circulating the relevant technical know-how was secured by several things. First, a range of instrumental appliances were specially modified for carrying out basic astronomical computations. Second, a few essential sets of rules and relevant observational tables were circulated as regulatory handbooks to guide pilots. These enabled navigators with no more than a rudimentary grasp of astronomical principles to figure out their latitude by combining basic observation and elementary trigonometry. Third, and most important, was the systematic training mariners underwent. Taken together, these practices were intended to enable communication centers like Lisbon to manage from afar computational operations carried out at the other end of the world. Devices, documents, and drilled people, as one student of Portugal's methods of "long-distance control" puts it, were fundamental to the circulation of knowledge and practice more generally. For many metropolitan natural philosophers, the model observational emissary was a well-drilled worker whose senses could be trusted because they had been coached at home for performance abroad.

As the sixteenth century wore on, we can see tactics of this class being put into effect at various European centers to deliver distant but dependable data. In Basel, Venice, and Paris, a suite of texts intended to instruct travelers in the arts of geographical observation made their appearance. Frequently designed to fit in with the scheme of learning advanced by the French anti-Aristotelian logician Petrus Ramus, such documents provided exemplary sketches of regional description. Significantly, they also included questionnaires directing travelers to those matters of greatest observational consequence. Just *what* should be observed and *how* such observations should be taken were rehearsed in detail. In the *Tabula Peregrinationis* of Hugo Blotius, which dates from about 1570, over one hundred questions were presented to enable a visitor to accurately record the features of any city. In such ways the eyes of the distant traveler could be disciplined to attend to matters domestically deemed significant. And at the same time, it was hoped, fickle memory would give way to foolproof method. Thus when Robert Boyle's anonymous guide for travelers appeared posthumously in the first volume of the Royal Society's journal under the title *General Heads for the Natural History of a Country, Great or Small,* it was the continuation of a long-established tradition of "methodizing" travel. And the same was true a century or so later when the Swedish naturalist Carolus Linnaeus compiled several texts containing systematic instructions for gathering medical and scientific information by explorers. Indeed, circulated queries could also be used for accumulating information within one's own country. In late eighteenth-century Scotland, for example, Thomas Pennant circulated a list of twenty-seven questions to "Gentlemen and Clergy" in remote parishes about local antiquities and natural history. Such people had the social standing to be relied on to act truthfully, and Pennant's questionnaire would direct their attention to the sort of information he wanted to compile. In this way virtual witnessing could be achieved.

Yet none of these tactics delivered certainty. Efforts to establish with precision the longitudinal position of various sites in the New World during the early 1570s using a circulated set of queries failed miserably. Incomprehension, misunderstanding, transcription errors, irrelevant information, and a host of other things persistently got in the way. But there was no other means of disciplining sources; the only option was to simplify the guidelines and press on.

Still, the general strategy persisted. In 1854 the Royal Geographical Society of London brought out the first edition of its *Hints to Travellers,* a handbook for scientific explorers that appeared in one new edition after another for decades. Here again the underlying concern was to resolve the problems of field observation by providing advice on essential equipment, instruction in instrument management, and a series of other "hints for collecting geographical information" (fig. 31). But achieving a regulated system of geographical inspection proved elusive. Experienced travelers differed in the observational details they reported, in what they took to be appropriate accuracy, and even on which pieces of equipment were the most fitting for expeditionary purposes. So over the years the precise mechanisms that were sought to deliver credibility changed. That the senses

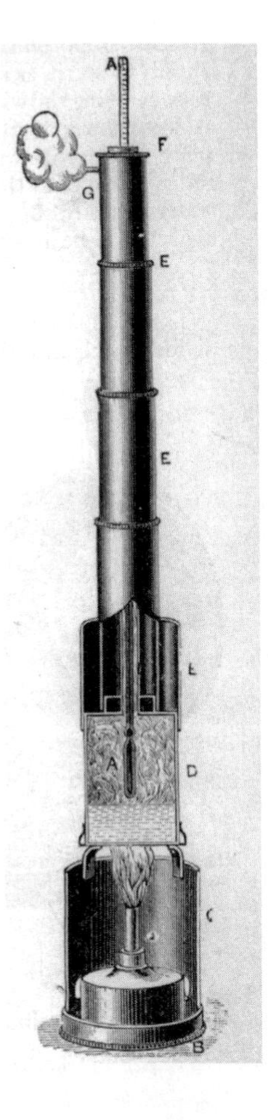

31. An illustration from the Royal Geographical Society's Hints to Travellers *of a hypsometer or boiling-point apparatus, an instrument used to determine height. In the attempt to regulate scientific travel, the Royal Geographical Society considered training observers in the use of such equipment to be essential for acquiring reliable scientific knowledge of distant places.*

of travelers needed educating was one thing; just how to achieve this end was quite another. Darwin's cousin Francis Galton, who exerted a large influence on the second, third, and fourth editions of the work, for example, had felt the need to provide advice in his 1855 book *The Art of Travel* on the necessity for expedition leaders to display self-discipline and on how they should conduct themselves with both fellow Europeans and "natives." Trustworthiness in personal character was all of a piece with trustworthiness in scientific reporting. Acquiring distant knowledge depended no less on moral fiber than on technical competence.

In some cases, too, the disciplining of the senses and the deprivation of the body were taken as mutually confirming. That an explorer's body had undergone the rigors of hardship in forbidding surroundings—literally bearing the marks of an alien environment—was considered the insignia of a trustworthy testimony. The demonstration of *moral* courage through its inscription on the explorer's *flesh* was thus taken as a token of *cognitive* reliability. For scientific travelers, the mental, the moral, and the material were routinely merged. Take, for example, the controversy surrounding who should be credited with the distinction of being the first European to set foot in the African city of Timbuctoo during the 1820s. In commenting on the matter, John Barrow, permanent secretary to the Admiralty, contested the claim that the glory should go to a young Frenchman René Caillié. Instead, he urged that the honor belonged to the Scottish soldier Alexander Gordon Laing. What is interesting in this case were the moral elements that Barrow introduced into the debate in his quest to establish trust. Because Caillié had entered Timbuctoo disguised as a destitute Arab and had posed as a Muslim convert, Barrow insisted that he had proceeded throughout by subterfuge and deception. "One who is thus ready at invention at first starting," Barrow sneered, "could find no difficulty in improving as he proceeded." Such ungentlemanly behavior stood in marked contrast to the nobility of Laing, who had heroically and painfully acquired his geographical knowledge. He had "practised no deception." Laing himself reported in literally agonizing detail the twenty-four wounds he had sustained in his clashes with Tuareg brigands, including multiple saber slashes to the head, left temple, and right arm, a variety of fractures, and a musket ball in the hip. Because credibility was invested in the authority of the person, the moral economy of wounds assumed great importance in calibrating trust. To Barrow, Laing was the epitome of self-sacrificial virtue, and the injuries he had sustained were nothing less than the signs of truth imprinted in the flesh.

The factual claims that circled the globe with scientific travelers, it is clear, raised critical questions about credibility. Just who could be believed and whose word could be trusted? One way of addressing this problem was to invest confidence in those explorers whose senses had been disciplined by technical, intellectual, and moral training. However this was to be achieved, the circulation of scientific knowledge was an inescapably *social* affair involving judgments about people. But trust was not solely located in human beings and their sensory apparatus. It also resided in a range of documentary registers that could travel independently of the people who produced them. Among these, the map as a device of translating knowledge from one space to another looms large.

MAPPING TERRITORY

The map has widely been regarded as an efficient and reliable way of bringing the world home. Whether in the ancient Roman world of Ptolemy, the ninth-century Chinese world of Li Chi-fu, or the Islamic world of al-Balkī, maps have been taken as graphic descriptions of the world. In the wake of the European age of reconnaissance, maps proliferated, and their scientific status was further reinforced during the Enlightenment. Of course later cartographers would routinely disparage earlier mapping enterprises as primitive, erroneous, even monstrous. Medieval world maps, for example, were subsequently castigated as unscientific and relegated to the pit of "complete futility." But the idea that truths about distant realms can be known through cartographic endeavor has been widely promoted. The Mediterranean sea charts of the late middle ages—portolans, as they are known—for example, progressively delivered more and more accurate depictions of coastlines that made them an indispensable nautical tool. Or again, the world maps of such Renaissance mapmakers as Ortelius and Mercator bear an astonishing resemblance to the shape of the world as we now recognize it. The New Atlas produced by the Dutch cartographer Joan Blaeu was widely regarded, and rhetorically boosted, as the major symbol of the Renaissance spirit of free inquiry, liberated from the shackles of the past. In a culture more and more thirsty for visual geography and faithful draftsmanship, such cartographic productions soon became valued as units of intellectual, commercial, and aesthetic currency.

Besides the topographic mapping of terrain, a remarkable range of other items has been reduced to cartographic form. By the eighteenth century, expeditionary endeavors had delivered charts of magnetic deviation, atmospheric circulation, and ocean currents. Soon maps of linguistic families and climatic patterns, the distribution of animal and botanical species, poverty and disease, mammal migration, and religious affiliation were also available. The list is enormous. And it would now include maps of the AIDS virus, the human genome, and the brain. Whether for the pragmatic purposes of navigation or the cognitive interests of scientific inquiry, people rely on the map as an accurate representation of the world under scrutiny. The power that maps exert in society is bound up with the impression of exactitude and precision that they convey. Their capacity to move information across the globe with remarkable ease is no less important. Data collected at the ends of the earth can readily be transported in manageable map form to central locations and then assembled, analyzed, and collated to disclose hidden patterns. With these powerful qualities it is not surprising that the map has been used as an analogy for scientific theory itself. For all these reasons, maps are repositories of trust. And yet as we now begin to scrutinize the idea of cartographic accuracy, to unpack just what is involved in map mobility, and to probe beneath its image of scientific neutrality, the map's trustworthy innocence begins to dissolve.

The idea that the map is a straightforward representation of reality is deceptively simple. It is this taken-for-granted assumption that makes it such a powerful device of persuasion and a source of cultural power. But once we start to dissect cartographic practice, the presumption that the map is a mirror begins to be exposed as an act of faith. For a start, every map is a controlled fiction. Because the earth is a globe, representing it on a flat surface requires using a projection to transform a three-dimensional sphere into a two-dimensional surface. All projections necessarily distort the map in one way or another; for example, either the distances or the shapes may be correctly portrayed, but not both. Mercator's famous map projection of 1569, which has delivered an image of the world we are all familiar with, conforms to the shape of the continents but not their relative size. Mercator's maps were deliberate manipulations for navigational purposes in order to preserve uniform compass direction.

Every map is a distortion in a second sense. It is a simplification of the reality it purports to depict. If it included everything, it would not be a map at all. Making a completely comprehensive map would require plotting at a scale of a mile to the mile—something that, as one of Lewis Carroll's imagined characters quipped, "would cover the whole country, and shut out the sunlight!" In such circumstances it would be wiser, he reckoned, "to use the country itself, as its own map." Evidently every map omits something, and these exclusions or "silences," as they have been called, can be immensely significant whether they arise from suppression or selection.

One or two examples will make clear just how powerful a tool cartographic erasure can be. When Columbus and his cartographic successors began the task of reducing the new world to maps, they effectively dissolved the local geography of native peoples. They renamed features and obliterated Indian denominations; they inserted images of exotic creatures and monstrous races; they erased all traces of the indigenous knowledge the surveyors had relied on. By disregarding patterns of tribal settlement, by conveying the impression of unoccupied lands ready for European occupancy, by employing European sign conventions for geographical features, and by importing coats of arms, royal insignia, flags, and religious emblems, they dissolved native geography. Juan de la Cosa's world map of about 1500, for example, displays European flags on a much enlarged Brazil whose coastline is studded with place-names commemorating shrines to the Virgin in Castile, Catalonia, and Italy. Mercator's double-cordiform (heart-shaped) projection of the world constructed in 1538 reveals a South America vacant save for Amazonian cannibals and Patagonian giants—a cartographic summons to further exploitation. As for colonial North America, the territories of the native peoples were effectively wiped out by European cartographers who operated with entirely different ideas about land and property and simply drew their own lines right across Indian nations. Here the moral politics of lines on paper dramatically revealed itself, for through these inscriptions local peoples were silenced.

What is true of the early mapping of the Americas, with its absences and omissions, was replicated elsewhere. When James Cook named well over one hundred Australian capes, bays, and isles, frequently using the names of European naturalists, he at once effaced local designations and brought those spaces into European circulation for the first time. In nineteenth-century India, the use of Western measurements and surveying techniques had the effect of reducing the subcontinent to manageable form by delivering a more systematic, rationalized representation than had hitherto been available. Imperial geodesy sought to make India over in the image of Britain—a space scientifically measured, systematically archived, and coherently regulated. Not surprisingly, local people often resisted, fearing—with much justice—that the surveyor would too soon be followed by the taxman. But their conceptions of space found no place in the survey sheets that geographically constructed India in the century or so after 1765.

Much the same was true of George Vancouver's surveying ventures in the Pacific Northwest. Following instructions from the Home Office in London, his brief was to produce a "compleat" geography of the coast. But that totality certainly did not include any evidence of native occupation. Though scientifically constructed using lunar observations, chronometers, flat Gunter's chains with logarithmic lines, sextants, and the like, his chart recorded only those things he himself deemed significant and that the Admiralty approved of. The outcome was the creation for British diplomats of "an anticipative geography," a sort of cartographic silhouette with blank spaces inviting imperial interrogation.

The map's use of projection and simplification render it a useful fiction. Its capacity to erase and reinscribe makes it a powerful fabrication. And with such historical signals, we surely have good reason to suspect that what we might call "cerebral silences" and "genomic erasures" are every bit as likely to be features of the mappings of the human brain and genome that have been so enthusiastically greeted in rather more recent days.

What further contributes to cartographic potency is the map's remarkably *mobile* character and its capacity to carry vast amounts of observational data from continent to continent, from periphery to core, from point of collection to center of calculation—and back again. Territory cannot migrate across the globe, but marks on paper certainly can. Something of how the power of maps is bound up with their mobility begins to become clear when we reflect on the activities of the eighteenth-century French navigator Jean-François de La Pérouse. Scouting the Pacific in the service of Louis XVI, La Pérouse had come upon Sakhalin, an island north of Japan in the Sea of Okhotsk, and sought to determine from local people whether it truly was an island or a peninsula. To his surprise, they displayed remarkable geographical awareness and navigational knowledge, and they conveyed their understanding by drawing a map on the sand. The difference between the islanders and the visitors clearly was not in cartographic ability or territorial comprehension; rather, it lay in the Europeans' capacity to carry home in written form, in graphic inscription—in short, in a map—information collected thousands of miles away. Local geographical knowledge expressed in scratches on the sand vanished with wind and waves. Its European counterpart remained in material form and circulated around the planet. The political power such a technique conferred on the West was immense as it flowed through the medium of documentary recording and networks of information retrieval. To be sure, the traces on the sand were not delivered to La Pérouse and his men without discussion, negotiation, and interpretation. The encounter between "local natives" and "visiting savants" was complex, involving gestures, gifts, translation, and, to one degree or another, trust. But the material marks on movable paper—the residue of that distant rendezvous—were the vehicle by which the voyagers brought the faraway back home.

What makes map mobility possible, of course, is that maps are transmitted using a code that can be deciphered by recipients. This is what conveys credibility. Drawings that do not adopt recognizable modes of cartographic representation find themselves discredited as primitive, aboriginal, unscientific, or some such. It is only when a visual language employs conventional rules that have been acquired by practitioners that it can begin to move meaning over long distances. When this happens, the local conditions of a map's making are hidden and the map travels with remarkable efficiency. Such an uncoupling of text and context gives the impression that the map discloses universal truth. Its content seems disconnected from any local circumstance or particular social structure. Map knowledge seems nonindexical; that is, its truth does not depend on any contextual factors. Thus it is worthy of our trust. But in fact a good deal of mapmaking is altogether customary. The lines on maps that we are all familiar with and that seem so "natural," such as the grid depicting lines of latitude and longitude, are entirely conventional. It was, for example, at an international meeting in Washington in 1884 that it was *decided* that in future it would be assumed that the 0° line of longitude passed through Greenwich. More generally, because maps embody tacit rules of procedure and conventions of communication, they can operate only within a social group that understands their visual vocabulary.

Map mobility requires sign stability. And yet the precision and lucidity of signs on maps mask underlying instability. Perhaps the most obvious of these signs are the boundary lines between contending political powers. The seeming fixity and clarity of national territory that imperial administrators saw on the maps surveyors brought home with them cloaked the ambiguity and fluidity of the space that "underlay" its representation. For surveyors, the ideal boundary line, fixed and rooted, needed to be composed of points with significant meanings in several registers. The line was to be "historical" in the sense that it was rooted in long-standing tradition, "natural" in coinciding with landscape features, "accurate" as determined by astronomical readings, and "visible" so as to consolidate its significance. Deploying such prescriptive criteria, however, was never easy, as

Robert Schomburgk, Royal Geographical Society gold medalist and boundary commissioner in British Guiana during the early 1840s, amply discovered. While he himself affected a preference for the "natural boundary" in opposition to what he castigated as "imaginary lines," again and again he found himself mired in negotiations between the deliverances of history, nature, practicality, and his own hard-won experience. So much was this so that in his private manuscripts, boundary lines sometimes moved from one side of a river to the other and then on to some further watershed! Establishing points of reference in what has been aptly called the "merciless homogeneity" of the forest was so dizzying an experience that, having lost his way on one occasion, he ended up temporarily losing his mind. Besides, what Schomburgk thought of as natural markers on the southwest border with Brazil were regarded as decidedly *un*natural by the Brazilians. What is deemed "natural" is a cultural judgment.

Schomburgk's final map, which brought order to the tangled web of the points he plotted, erased all local traces of the contingent and circumstantial particulars of its making. But it transformed territory hitherto unknown to European powers into a space of determinate shape and size. Subsequently, in various disputes about territorial possession between Guyana and its neighbors, it is ironic that the very lines that brought the national territory into being found themselves contested by the entity they had created. Broadly similar maneuvers are readily discernible elsewhere. The provision of a cartographic delineation of the "geobody" of Thailand in the decades around 1900, for example, made possible an incongruous retrospective projection into history of a "Thailand" that did not exist until the map constructed it. In this case too, the making of national identity was intimately bound up with the production of a cartographic image of territory. Cartographers, it is clear, manufacture power by their capacity to create what one observer has fittingly called "a spatial panopticon."

In the light of these cartographic performances, the analogy that is commonly drawn between the construction of maps and the devising of scientific theories is particularly telling. Michael Polanyi, for instance, spoke of theory as "a kind of map extended over space and time." Thomas Kuhn considered that "paradigms"—traditions of scientific inquiry—provided practitioners with "a map" and with "directions essential for map-making." If indeed this comparison is well founded, it can only be on the understanding that both maps *and* scientific models reflect the local conditions of their making and act to construct the very entities they purport to disclose. And this is surely all the more so where scientific theory is cartographically constituted. Three cases will illustrate something of these maneuvers: the use of the isoline as a technique of graphic representation; the delineation of faunal boundaries by Darwin and Wallace; and Roderick Murchison's naming of geological strata. In each case the scientific maps produced were, in important respects, cultural productions. That they convey every impression of impartiality and neutrality, and thereby inspire trust in their objectivity, is itself a mark of the power of cartographic discourse to present as natural what is culturally constructed.

Although it had been in use at an earlier time, the isoline as a cartographic tool was brought to prominence by Alexander von Humboldt in 1817 when he published his major findings on the global distribution of heat. In doing so he used lines that connected points of equal thermal value. By this device he could impose coherence on miscellaneous numerical data across space and make large amounts of information visual. He also coined such new terms as isothermals (for heat) and isodynamics (for magnetic intensity). It was the self-conscious creation of what has been called an "isoworld." But this literal "worldview" was more than a cartographic composite of instrument readings. It was intended to convey to Humboldt's audiences the unity of the natural order, the connected nature of things, the "cooperation of physical forces." Measurement and mapping became the means by which Humboldt sought, through global physics, to represent the world as an organic whole. Humboldt's grand project was as much a work of aesthetic sensibility as of computational cartography. The isoline enabled the student of nature to penetrate surface chaos to discern the inherent harmony beneath. Within a few years this representational contrivance was being used to construct a vast range of isomaps—of rainfall, temperature, cloud cover, ocean depths, and much else. The popularity of these procedures contributed massively to the internationalizing of science in the nineteenth century. But it also brought isolated observations into global frameworks and revealed hitherto hidden distribution patterns. The isoline was at once an exercise in the development of a new geopolitics of science and a massive investment of trust in the power of science to cross physical, cultural, and language barriers.

For both Charles Darwin and Alfred Russel Wallace, the idea of zoogeographic regions with definite boundaries was crucial to their evolutionary theorizing. By plotting demarcation lines on maps, the range of global plant and animal life could be framed and fixed. The resulting maps were theoretically stimulating. The patterns they disclosed prompted questions about species origin and migration and contributed to the visualizing of evolutionary theory. As Wallace himself famously put it, "Every species has come into existence coincident both in space and time with a pre-existing closely allied species." Yet the boundaries were far from self-evident. Darwin, for example, wavered on the number of worldwide zoological regions, eventually plumping in 1844 for five. And famously, Wallace literally made his mark by his drawing of the "Wallace line," which constructed the border between Indo-Malayan and Austro-Malayan fauna. When it appeared in the *Proceedings of the Royal Geographical Society of London* for 1863, the line efficiently delivered data from afar and effectively imposed clarity on distant complexity (fig. 32). The map made visible the world of living things as Wallace wanted it to be seen. But there is good reason to suppose that the impulse toward charting borders was rooted in Wallace's obsession with human ethnicity. And this can be traced back to his early experience in rural Wales, where in the late 1830s he had been employed as a land surveyor. Here, as he witnessed the grim realities of rural poverty and followed the ancestral boundaries of Celtic peoples, he came to appreciate the power of ethnographic cartography. Half a world away in the Malay Archipelago, he again surveyed racial geography and constructed an ethnological line just a few hundred miles east of his cele-

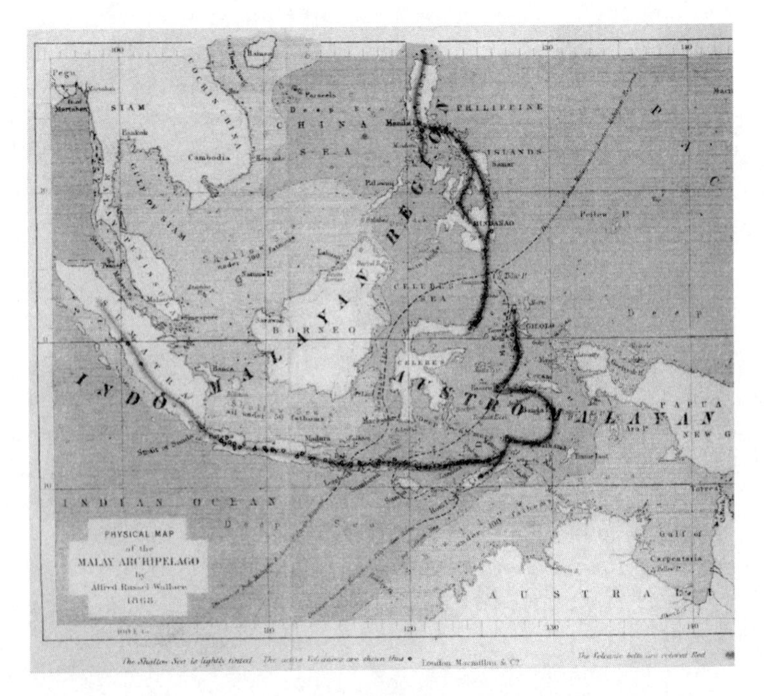

32. The faunal and racial boundaries of the Malay Archipelago, by Alfred Russel Wallace. The faunal boundary was first published in 1863. A year later Wallace constructed a racial boundary that he insisted was almost as well defined as its zoological counterpart.

brated zoological line. For Wallace, human geography and animal geography were always intimately intertwined. And maps became a strategic rhetorical device through which he could conjure into view both the social and the zoological facts his theories sought to explain.

Similarly, by placing names on map sheets, scientific entities of various kinds have been brought into cultural currency. Through the use of labels, the Victorian geologist Roderick Murchison brought unknown lands in Africa under the sway of geological terminology. He had elucidated the Silurian strata and was determined to extend its jurisdiction across the face of the earth. His doing so admitted the

African landscape into the international geological conversation and placed it under the authority of a Western scientific outlook. And as he enlarged his terminological kingdom, through orchestrating the Royal Geographical Society's expeditions to the "dark continent" for twenty years or more, he resorted to imperial language to describe how his taxonomy "invaded" continents, "enlisted recruits," and engaged in "the field of battle" much like the ancient Romano-British tribe for which the Silurian strata were named. Not surprisingly, he was every bit as concerned to advance Britain's imperial interests as he was to extend the empire of Siluria. His terms circled the globe like the tentacles of British imperialism. The process of naming thus turned out to be an exercise in colonial expansionism. For Murchison, cartography provided rulers with administrative apparatus and imperial instruments as well as conceptual devices for comprehending and governing the world.

When maps carve the world up into seemingly coherent zones, when they name places and natural objects, when they categorize creatures and commodities, when they claim to bridge the gap between near and far, they invite our trust. They can lure us into thinking we are witnessing the world. But they cannot, by their very nature, replicate the world. Maps are not facsimiles of the planet. And the extent to which we think they are demonstrates the influence the cartographic image has over us.

PICTURING THE UNFAMILIAR

If maps cannot accurately reproduce the world, perhaps pictures can more reliably bring the remote within reach. Writing in the *Art Journal* for 1860, one observer insisted that because the photograph could not deceive, "we know that what we see must be TRUE. So guided, therefore, we can travel over all countries of the world, without moving a yard from our own firesides." Photographs, of course, were just the most recent pictorial strategy for inspiring confidence in the reliability of testimony. They became surrogates for firsthand witnessing. And they seemed to have considerable advantages over earlier artistic representations of nature, which were notoriously tricky to credit. When travelers brought back illustrations of strange creatures and unfamiliar plants, doubts about their trustworthiness were not hard to raise. And a variety of strategies were put in place in the effort to provide credibility.

Chief among these, as James Cook recognized, was the use of professionally trained artists, and he therefore took illustrators on all his voyages. Perhaps aware that natural history pictures had potentially different audiences, he used different artists to appeal to aesthetic connoisseurs and to natural history savants. Nevertheless, what Cook was after was an empirical *style* of pictorial representation that was more in keeping with scientific thinking than with artistic convention. Uncomplicated, restrained, unadorned: these were to be the hallmarks of scientific illustration in the Cook mold. For these qualities conveyed the sense that an artist had carefully scrutinized a real specimen, and had scrutinized it up close. Simplicity and precision had the ring of truth; ornamentation and decoration did not. Cook's illustrators thus played their part in the lengthy historical shift from a classical toward a natural style.

And yet even though this move itself resonated with a contemporary British aversion to French frippery in matters of artistic preference, there remained a tension between the call of taste and the demand for precision. Joseph Banks's natural history painters, for example, *did* at times devote their energies to romantic topics like grottoes, exotic rituals, and so on, because these suited the fashionable baroque tastes of some contemporaries. Moreover, even when they made accurate depictions of native peoples, like those instances of documentary realism produced by Alexander Buchan (a landscape painter Banks enlisted), it just was very hard to bring an undoctored account of them before the public. Engravers *would* dress up the original painting to bring it in line with their own predilections. John Hawkesworth, for example, time and again allowed his enthusiasm for primitivism to come through in the illustrations he used, portraying those "noble savages" as modern exemplars of austere virtuousness. On the one hand, he told the readers of his 1773 *Voyages* that his account of the Patagonians was dependable because it was corroborated by "the concurrent testimony" of several naval "Gentlemen of unquestionable veracity" who had seen, conversed with, and measured these peoples. At the same time, the natives of Tierra del Fuego whom Buchan depicted as living a life of misery found themselves transformed, in the Hawkesworth engraving, into exemplars of primeval dignity (figs. 33 and 34). The squalid had given way to the graceful. Evidently breaking free from representational custom was difficult even if explorer-artists strained to allow their art to be structured by reference to the world itself rather than by normative notions of the picturesque or the whisperings of a distant divine.

Nevertheless, pictorial illustrations were often promoted as reliable testimony and were thus accompanied by textual invitations to trust. It was said that the information on native costumes that found expression in the drawings of John Webber, who had accompanied Cook, could be replied on because they were done on the spot. Firsthand inspection, proficient draftsmanship, a documentary style, and disciplined eye-hand coordination were taken as security. It was for just such reasons that Hawkesworth's adulterated engravings called forth criticism from some scientific readers who felt that artistic ornamentation had too fully triumphed over nature's simplicity. Scientific illustration was thus an arena where battles were fought over who could be trusted to deliver knowledge of distant realms. The designation "reliable" had to be won. Artistic integrity was a social achievement. And establishments like Kew Gardens acted to stabilize it. By providing accredited images against which new illustrations could be matched, such institutions acted to calibrate trustworthiness. They had the power of adjudication because they had acquired untouchable cultural authority. In turn, pictures themselves exerted immense influence as they circled the globe and brought imagined realms before the mind's eye. The very idea of a Pacific world owed much to the artist's powers of evocation. But could they be trusted? That doubt still nagged.

And so the photograph was welcomed in glowing terms. By mechanical reproduction, photography could furnish a verisimilitude

33. "Inhabitants of the Island of Terra del Fuego in Their Hut," painted by Alexander Buchan in 1769. Buchan's sketch confirmed Captain James Cook's belief that the Fuegians were a miserable set of people.

beyond the technical competence of any artist. The Scottish physicist David Brewster, for example, sang its educational praises because it delivered "accurate representations." To him, teaching "through the eye" was the key to scientific instruction. Humboldt too was wildly enthusiastic about the documentary value of Louis Daguerre's invention. For in photography Humboldt saw the possibility of achieving his aims. When he mused, in a letter to an English friend, that "Daguerre is my Chimborazo," he was reflecting on that moment at the foot of the highest peak in the Ecuadorean Andes when he had glimpsed a holistic vision of the natural order. Photographic reproduction could replicate that ecstatic visual experience. Besides all this, photography (and particularly stereoscopic photography) offered the possibility of vicarious travel. As a review in the *Art Journal* for 1858 put it, photography presented "only the plain unvarnished truth; the actual is absolutely before us."

34. "A View of the Indians of Terra del Fuego in their Hut," an engraving by Francesco Bartolozzi that appeared in John Hawkesworth's An Account of the Voyages Undertaken by the Order of His Present Majesty for Making Discoveries in the South Hemisphere *(London 1773). This was based on Giovanni Cipriani's altered engraving of Buchan's drawing (fig. 33) and shows how Buchan's miserable Fuegians were transformed into people of primitive grace.*

Given such enthusiasm, it is not surprising that photography was embraced as a trustworthy means of overcoming geography. And it was soon put to use in a variety of scientific endeavors—astronomy, anthropology, medicine, meteorology, and geography itself. Photographic images could sidestep the problems of unreliable witnesses, cartographic silences, artistic embellishments, and the like. The welcoming of photography as an "expert witness" at the Royal Meteorological Society is a case in point. Because the development of meteorology required collecting data from a widespread network of spectators, practitioners embraced photography as a reliable, ever-present, and untiring observer that could catch and preserve things the naked eye could not even detect. Photography promised to ensure the very trustworthiness that was so hard to establish from the reports of lay eyewitnesses. It could act as a sieve to separate fact from fiction, information from imagination. For after all, such phenomena as lightning continued to be shrouded in myths about thunderbolts and the like. Thus Arthur Clayden, a fellow of the Meteorological Society aspired to build up "a great army" of observers equipped with cameras. Thereby meteorological data could be recorded accurately, transmitted over long distances, and compared—the very things it was extraordinarily difficult to do with human testimony.

Or could it? As it turned out, photographic evidence created as many problems as it solved. For photography was an artistic craft. Not only were lengthy preparations and appropriate apparatus required, but it was often unclear whether the image recorded was the outcome of human error, an artifact of the instruments used, or a genuine mirror of nature. Resolving these questions necessarily required judgment. Moreover, it was often hard to get photographers to resist the picturesque appeal of lightning photography. One contemporary meteorologist complained that keeping an eye on the "pictorial effects" too often prevented photographers from "inserting anything so ungainly as a yard measure." Meteorological photographs clearly did not overcome the problem of trust; they simply extended the scope of dubiety.

If advocates of weather photography found it difficult to secure trust, nineteenth-century travel photography demonstrates this species of difficulty in excelsis. It was very largely through published works of photographic illustration, intended to portray the glories of Greece or the mysteries of Egypt, that "imagined geographies" of the world circulated widely. Supposedly reproducing the *real* world, travel photographs constructed an *imagined* world through the lens of the camera. The visual inventories that travelers brought back reflected not only sponsorship—whether from government, science, or commerce—but also the limitations of the techniques themselves. The cumbersome nature of the equipment, together with the bulky supplies that were needed, frequently dictated the locations from which a photograph could be taken. Besides, in some cases human figures were erased on account of long exposures, or were deliberately inserted into the scene to humanize it, or were introduced simply to provide scale. Ultimately, as one commentator has tellingly put it, travel photography "reduced sites to sights" by privileging visual knowledge over the multisensory experience of foreign travel with its sounds, smells, and sensations.

Still, it was very largely through photography that distant spaces were brought before domestic audiences, both popular and scholarly. Sir Halford Mackinder, geographer and member of Parliament from 1910 until 1922, saw in photography the means of advancing his project of educating citizens in imperial geography. Through his key role in the Colonial Office Visual Instruction Committee—a body set up to present the sights of empire to mass audiences in Britain—Mackinder instructed the committee's photographer in exactly what should be recorded for lantern slides. No less popular were the photographic displays of the *National Geographic*. Here idealized photographs of non-Western peoples inhabiting timeless worlds were repeatedly brought before the public.

At the more scholarly end of the spectrum, documentary photography has been widely used in both physical and cultural anthropology to record variations in the human physical form and to chronicle ethnic customs and practices. The technique of producing facial composites (in which a "typical" head form was photographically distilled from many exemplars), together with straightforward shots of "representative" individuals, confirmed for anthropologists that different races possessed stereotypical features (fig. 35). Significantly, composite profiles were often mapped straight onto supposed mental and moral quality. Darwin's cousin Francis Galton used the method to construct what he took to be the typical criminal physiognomy. As for capturing ethnic customs, careful staging was often required to secure an image that looked sufficiently natural. In all these cases, the photograph has acted to *construct* the identities of the places and peoples it seemed to represent.

Photographs, then, like paintings and maps, have always been

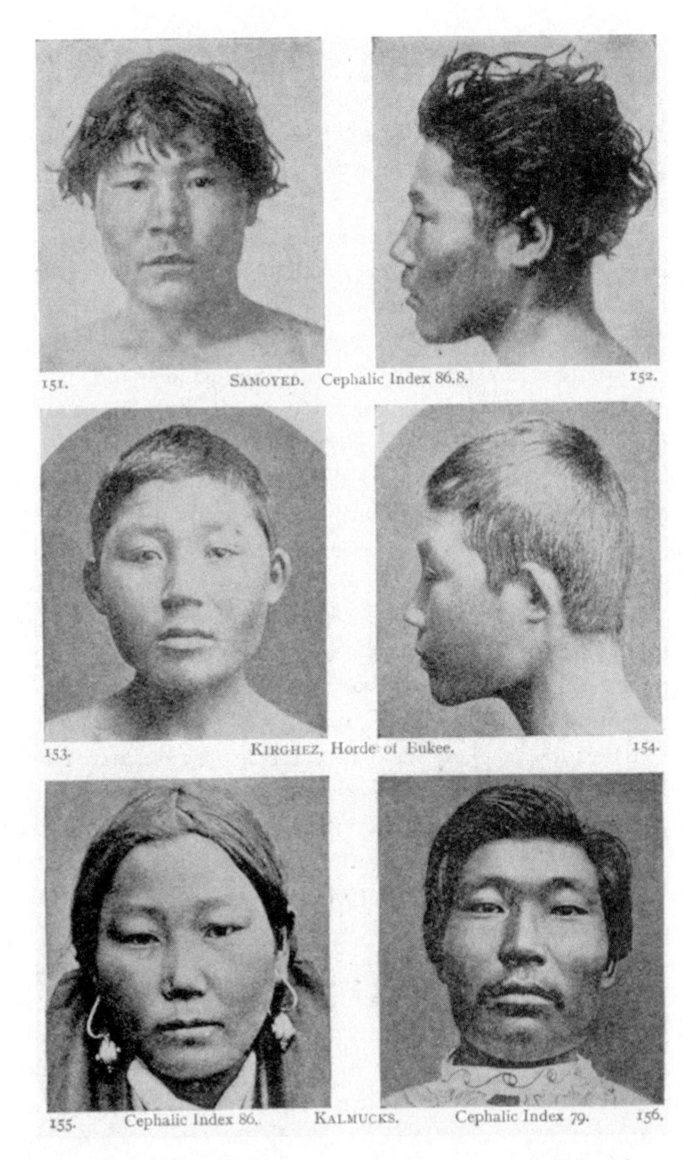

35. Photographic illustrations of typical racial types, in this case "Mongol types," from William Z. Ripley's Races of Europe *(1899). Photography, with its sense of scientific objectivity and realism, was used to construct senses of racial identity.*

the work of situated observers. And their stock of pictures is as much the compound product of patrons' desires, audience appetite, artistic taste, and technical possibility as of the realities of nature. Adopted as a sure-shot means of overcoming distance and guaranteeing trust, photography only served to rerun these selfsame problems through a different medium. At the same time, the very fact that photographs were exploited in radically different sorts of ideological campaign—for imperial surveillance by administrators to promote the interests of empire, for social critique by radical reformers, for anticolonial resistance by missionaries who recorded the aftermath of military brutality, for surrogate voyaging—recalls attention to the *rhetorical* character of the photograph and to the artful nature of its seeming objectivity and neutrality.

Gathering the World Together

The disciplining of observers' senses, the translation of data into cartographic form, and the use of photographic technology as a recording device were just some of the ways efforts were made to obtain dependable knowledge of distant phenomena. Such tactics were intended to obliterate, as far as possible, the space between near and far, here and there, presence and absence. But as we have seen, credibility could never be secured without recourse to *judgments* about the trustworthiness of people and their performances. At the same time, the whole point of seeking to establish techniques of trust was to make knowledge mobile so that it could circulate from acquisition points to assemblage spaces where collation, comparison, and recombination could be carried out. Collecting data was simply the first step in gathering the world together.

The "centers of calculation," as they have been dubbed, to which the produce from global data harvests returned enjoyed immense power. These spaces—like the museum and the botanical garden we considered in chapter 2—were nodal points in the flow of information and controlled widespread networks of communication. By so doing they also held the power to shape the way the world was put together, not least by their role in condensing the earth to the scale of a chart or an index or a catalog. From the disparate materials they acquired—specimens, maps, images, records—these centers forged a global panorama. On paper, in cabinets, and on tables, articles from different locations and times come together to share a common space. Objects collected years and miles apart find themselves united in new combinations. Here samples become signs, entities become numbers, physical features become cartographic lines.

Perhaps the earliest of these compilation sites was the Casa de la Contratación in early sixteenth-century Seville. This bureaucratic "knowledge space" was essentially a board of trade charged with the task of managing Spanish commerce with the East Indies and the New World. Hydrographic control was vital to the enterprise. And so the Casa's twin-pronged mission was to retain a monopoly on cartographic knowledge and to regularize the information that seafarers brought home. This early exercise in data handling required careful management so that local knowledge could be merged into more general cognitive systems. The master chart or template map (the *padron real,* as it was known) was the outcome. It was fundamentally an aggregate nautical chart, combining maps that themselves were the product of loose compilation practices. The Beccari chart of 1403, for example, incorporated several scales. In due course the state encouraged more standardization, and tables were circulated for calculating distances, determining latitude, and the like. The Casa—which came into being in 1503—had the task of achieving cartographic integration by regularizing the ad hoc.

Similar roles were later performed by institutions like Kew Gardens and the British Museum, which in turn built up worldwide networks of satellites. The imperial complexion of these sites has already attracted our attention. Here it is enough to recall that during the Victorian period, Kew was the pivot of a colonial complex of gardens in Calcutta, Jamaica, Singapore, St. Vincent, and Mauritius that it orchestrated from its metropolitan vantage point (fig. 36). No wonder the Colonial Office, the Foreign Office, and the India Office were so

36. A lithograph of a drawing by Lansdown Guilding of the botanical garden on the island of St. Vincent (1825). Such colonial gardens promoted the global circulation of plant specimens and thus the advancement of imperial botany on behalf of Kew Gardens.

dependent on Kew's botanical expertise. Indeed, the Foreign Office reminded the Kew director in 1891 that "a proper knowledge of the Flora of Tropical Africa would do much to aid the development of the territories over which this country has recently acquired our influence." Plainly, scientific circulation was as necessary to the intellectual composition of the world as to the imperial conquest of the globe. For the arteries through which botanical information flowed were also those through which imperial power coursed.

Such centers, of course, did not have to be institutions of national scale. In the case of eighteenth-century botany, the home of Joseph Banks in London acted as a channel through which passed ideas and illustrations, specimens and samples. It became the hub in a geography of dispersal; cognitive and material items gathered during Cook's three major voyages were dispatched from 32 Soho Square all across

Europe. At the same time, Banks gave specific instructions on how natural history research should be conducted. He always sought collectors who were careful observers, who wrote in "a good hand," and who were well versed in methods of gathering and drying specimens; and he always preferred bachelors without family responsibilities who had no aspirations to become gentlemen and who were content with servants' lodgings. These activities gave him a central role both in the construction of the world of botany and in furthering Britain's overseas imperial interests. For his detailed briefings also included the mandate to assess the suitability of new lands for settlement. Not surprisingly, Robert Hay, permanent undersecretary at the Colonial Office, considered Banks "the staunchest imperialist of the day."

Maps and specimens were not the only items to migrate to, and from, such cognitive assembly points. Centers of calculation have traded, perhaps even more commonly, in numbers. Measurable items of all kinds—quantities, dimensions, weights—also circled the globe in the form of recorded readings. They ended up in computational depots where the raw data were manipulated into statistical entities. Census bureaus, commercial enterprises, environmental agencies, pharmaceutical companies, life insurance agencies, and many more have all sought to combat the tyranny of distance by gathering information from widely dispersed sites and combining findings into comparable units.

The constructive capacity of such statistical agencies is of very considerable proportions. By compiling data from different *départements* on education, illegitimacy, prisons, health care, and mortality, the Statistical Society of Paris, which flourished during the mid-nineteenth century, was able to produce statistical indexes of well-being for the state. When translated into map form, such enterprises could classify cities, states, even continents, into a bipolar taxonomy of the sickly and the salubrious. They could divide space into the pathological and the wholesome. When the tabulations annually produced by the Bureau de Statistique revealed a small number of males in their twenties marrying women in their seventies, these cases were rapidly seized on by the statistical manipulators. The result? Otherwise disparate individuals found themselves constituting a group that in turn became the locus of scholarly interrogation. The range of questions to be asked of such a collective—or of any other demographic data set now inhabiting the inner reaches of on-line digitized information archives—is well nigh limitless. Do they share ethnic or class characteristics? Have they common psychosocial profiles? Do they occupy similar niches in the political and economic order? Do they display any distinct pattern of geographical distribution or religious affiliation? The cluster may even crystallize—say as "gerontophiles" or some other such neologism. In this way statistical manipulation displays its capacity to create social entities—and then to exert power over them.

Centers of calculation, however, could operate with any conviction only if the data they manipulated had been obtained in some systematic way. Whether in gathering observations, recording findings, charting results, or doing calculations, the haphazard, the irregular, and the capricious were the enemy of knowledge circulation. In this context the ideal of "precision" was decisive. By the eighteenth century, it already had become a cardinal virtue by which judgments could be made about observers. The presence of tables of measurements in Enlightenment travel accounts, for example, became the emblem of the serious scientific traveler. Precision disciplined mere curiosity and channeled its energies in a scientific and, as often as not, imperial direction. But what was needed to underwrite precision and to enable comparison and combination was standardization of both measurement and procedure, whether the aim was to regulate water supply, to test new medicines, or to compute life expectancy.

Take the matter of recording color in various field sciences. Because color memory is highly unreliable, what is known as the Munsell code—a manual used by paint manufacturers—has become a routine piece of field tackle. On it various shades are accorded a standard reference number. Once a number is assigned to a piece of soil, say, it easily travels in the fieldworker's notebook and can be combined with other approved codes in order to address scientific questions. It is because the code circulates that common color reference can be made. By it a tiny piece of earth can participate in a universal code. Through it the particular and the general come together.

Standardization, then, is the prerequisite for conquering space—the space between the field and the center of calculation, and the space between nature and language. Only by using agreed-on standards could knowledge be relieved of the burden of parochial judgment or fickle memory.

Achieving standardization was far from straightforward, however, and here what might be called "the polity of number" clearly manifests itself. In eighteenth-century Europe, for example, agricultural produce was measured using local weights. But these were hardly sufficient for trade farther afield. What was necessary to overcome mensural variation—since maintaining a region's own bushel measure was regarded as a symbol of liberty—was the power of the state to set standards and ensure equity. In nineteenth-century Britain, the push to establish uniform standards, intended to deliver universality, was shot through with controversy. Because economic, political, and scientific interests all had stakes in standardizing measurement units, the debates were predictably multifaceted. That the regulation of economic exchange and scientific intercourse was taken to be of immense significance is amply demonstrated by the range of bodies instituted to carry out metrological surveillance. The Statistical Department of the Board of Trade, the Factory Inspectorate, and the General Register Office, for example, all came into being in the 1830s. In the following decade both the British Association for the Advancement of Science and the Excise Laboratory sought mechanisms to deliver mensural purity.

As for standardization itself, debate raged over whether standard units were natural or conventional, divinely sanctioned or humanly generated. In rigorously championing the cause of the imperial yard over against the French meter, advocates variously argued that the yard had the backing of tradition, that it was appropriately related to the obvious standard distance of the earth's polar axis, that it gave long-standing expression to Britain's commercial superiority over the French, and that it enjoyed divine warrant by virtue of its direct connection with the dimensions of the Great Pyramid of Giza. The British yard simply must triumph over the atheistic mensural system of French republicans! In France, the defense of a "natural" system of measures was bound up with a revolutionary ardor for obliterating all traces of monarchical caprice. By determining that the meter should be one ten-millionth of a quarter arc of the great polar circle (that is, of the distance between the North Pole and the equator), the needs of both science and ideology could be met. Precision itself became a political virtue in France as revolutionaries worked to emancipate citizens from old pagan ways of keeping time and instituted a ten-hour day.

During the nineteenth century, standard ways of measuring increasingly conquered the local, though in fact such procedures amounted to the triumph of one set of local practices over others. National meteorological networks, for instance, came into being because consistent ways of measuring temperature, humidity, wind velocity, and barometric pressure meant that atmospheric data from locations across the earth could be assembled in central offices. Now data mountains could be sorted through, categorized, refined, and manipulated in all sorts of ways to produce correlations and weather predictions. Standardization, then, was designed to overcome distance and distrust and to promote circulation. For by employing tried and trusted impersonal methods, analysts procured data about places and people far removed from the direct gaze of the information monger. Only then could college administrators compare standardized grade scores from Boston and Seattle without any local knowledge of either region. Only then could measurements of air temperature and precipitation travel from a distant recording station or a weather ship to a national meteorological office. Only then could large transportation networks operate according to schedule. In all these, standardization is needed to triumph over the local, to gather the world together, and to reassemble it from standardized units of measurement.

* * *

The growth of scientific knowledge has been intimately bound up with geographical movement. Thoughts and theories have migrated across the earth. Machines and models have diffused from place to place. Information gathered on distant shores has crossed the oceans in minds and maps and manuals. Sketches and samples have brought the unseen before scientific eyes. In these and a hundred other ways, scientific knowledge has been expanded by circulation. And yet the securing of this enrichment has raised profound questions about how knowledge is acquired. For distance and doubt have always been close companions. Knowledge of the faraway depends on the reliability of absent witnesses. And this realization has prompted the instigation of a whole range of mechanisms to warrant credibility. Observers have been drilled; bodies have been disciplined; pictures have been painted; photographs have been taken; maps have been charted; measurements have been standardized. Just how successful these strategies were, however, was always a matter of judgment. Findings were always open to suspicion and negotiation. The irreducible reality of space, and of circulation in the growth of science, is thus a potent reminder that scientific knowing is an inescapably social phenomenon involving judgments about the integrity of people and their practices. Geography makes the scientific enterprise an inescapably moral undertaking.

Putting Science in Its Place

Like other elements of human culture, science is located. It takes place in highly specific venues; it shapes and is shaped by regional personality; it circles the globe in minds, on paper, as digitized data. For these reasons alone science is as conspicuous a feature of the world's geography as patterns of settlement, the distribution of resources, or the configuration of cultural landscapes. Yet bringing science within the domain of geographical scrutiny seems disquieting. It disturbs settled assumptions about the kind of enterprise science is supposed to be. It calls into question received wisdom about how scientific knowledge is acquired and stabilized. It complicates the taken-for-granted division between science, society, and nature. Taking seriously the geography of science positions the local at the center of scientific ways of knowing. It confirms that the authorized apportionment between "the natural order," "social context," and "scientific inquiry" is a rhetorical device that imposes clarity on ambiguity. It renders suspect the idea that there is some unified thing called "science." That imagined singularity is the product of a historical project to present "science" as floating transcendent and disembodied above the messiness of human affairs.

Our travels amply confirm, however, that science is not above culture; it is part of culture. Science does not transcend our particularities; it discloses them. Science is not a disembodied entity; it is incarnated in human beings. For all the rhetoric that science is independent of class, politics, gender, race, religion, and much else besides, we have seen something of the extent to which it bears the marks of these very particularities. Botanists do not shed their ethnicity when they engage in fieldwork. Chemists do not discard their gender when they walk into a biotechnology lab. Anthropologists do not set aside their politics when they map ethnic differences. Science is not some eternal essence slowly taking form in history; rather, it is a social practice grounded in concrete historical and geographical circumstances.

Probing the geographies of science, then, is a necessary corrective to deficient conceptions of "the scientific enterprise." In our journey we have considered three ways of "putting science in its place." We have visited sites where science has been practiced; we have witnessed the mutual making of scientific culture and regional identity; we have untangled some of the webs of scientific circuitry. The diversity of venues where scientific inquiry has been undertaken and the different cultural formations that characterize these spaces are remarkable. Sites of experimentation like the laboratory, where the impulse is to manipulate the natural order by experimental intervention so as to discern how independently variable factors behave in their natural state, are spaces that are differently constituted than those sites of exhibition, like the museum, where the goal is the accumulation of artifacts and their rearrangement for display. At the same time, both of these are different from spaces of expedition, where immediate experience of raw, *un*manipulated, nature is taken as an epistemic necessity. And science has been practiced in many other venues too—in public houses and princely courts, on ships' decks and stock farms, in coffeehouses and cathedrals. In different arenas different repertoires of practical rationality have been in operation, and forms of explanation, modes of practice, methods of justification, and traditions of inquiry central to one arena have been outlawed or marginalized in others. The *where* of scientific endeavor thus insinuates itself into science at all these levels.

At a different scale of analysis—the region—science has been marked by geographical circumstances too. In different European regional settings, forms of inquiry that later came to be described as "the Scientific Revolution" bore the stamp of their local arenas of engagement. In some cases a maritime culture was the chief engine power behind the cultivation of scientific pursuits; in some a courtly culture predominated; in others religious conviction was the molding agent; in yet others economic ambitions provided both impetus and constraint. Elsewhere, and also at the subregional scale, other circumstantial combinations were operative. Whichever was at work, those activities that came to be gathered under the designation "science" were rooted in the particularities of place. And it is therefore not surprising that in different locations, scientific theories were received in different ways. Whether it was Newton's mechanical philosophy, or Darwin's theory of evolution, or Einstein's relativity, the meaning and implications of these scientific conceptions were differently construed in different places.

Just how knowledge embedded in a particular location moves from its point of origin to general circulation, and thereby transcends locale, is an inherently spatial question and introduces a crucial dynamic to the geography of science. Rather than being understood simply as an inevitable consequence of a uniformly constant nature, the universality of science is the consequence of a variety of practices that have had to be put in place to guarantee reliable transmission. The disciplining of scientists' own senses, and those of the witnesses they called on, has been one strategy. The deployment of standardized measurement and the development of statistics are others. At the same time, maps and photographs have been used to overcome the tyranny of distance by bringing home reliable knowledge of the faraway. Because they can record distant phenomena in static form yet move across the earth, these are what Bruno Latour has aptly called "immutable mobiles." What we have discerned, however, is that all of these operations, though intended to eliminate distrust, underwrite testimony, and guarantee credibility, are shot through with ambiguity and uncertainty so that matters of judgment are inescapable. The successful spread of scientific knowledge across the globe is at least in part the outcome of a series of situated practices specifically aimed at achieving scientific ubiquity.

Site, region, and circulation, of course, do not circumscribe the bounds of geographical readings of science. In these concluding remarks, I want to take a rather high-altitude view of a couple of further extensions of the enterprise. The first is what I call "life geographies," or the spaces of biography. One route into this territory is the kind of biographical mapping that has been undertaken to come to grips with the controversy in nineteenth-century England over the Devonian strata. Just how the Devonian strata were brought into geological discourse by the gentlemanly specialists of the day becomes dramatically clearer when we map the London locations of a number of key players in the geological drama—Lyell, Sedgwick, Murchison, Darwin, de la Beche, Phillips, Greenough. Their close physical proximity discloses the ease with which they could meet with each other, both at their homes and at various learned societies, to pursue private conversations and engage in intellectual exchange. What becomes plain is that in the making of geological knowledge, physical location, social positioning, and cognitive authority were intricately interwoven. For there was a conspicuous overlap between the positional geography of the elite players in the Devonian controversy, based as they were in London, and what has been dubbed the "cognitive topography" of British geological expertise. This does not mean that they all agreed with each other on matters of interpretation; rather, they constituted the arena in which matters of theory and method were vigorously debated and settled.

What I mean by geographical biography, however, goes beyond this and takes as its cue the renewed interest in place that has arisen of late on account of a sense of disquiet at the fragmentation of modern culture. As some philosophers have insisted, until we undertake sustained reflection on what it means to be "in place," dislocation and disorientation will continue to characterize the human condition. The "self" has become increasingly fractured. Compared with the time when a person's whole life was narrowly circumscribed within a limited space or "station," nowadays all of us occupy an immense range of different sites. In these we act differently, adopt different personae, call on different linguistic repertoires, project different "selves." Hence we can plausibly say that someone is "a different person" at home, in the office, on the playing field, and so on. This is because we define ourselves by reference to the positions—the moral and social spaces—from which we speak. The "geography of social statuses and functions"—as one philosopher has written—provides the defining relations within which we construe ourselves. Morally and materially, *where* we are matters a good deal in trying to figure out *who* we are.

There are ramifications here for the writing of scientific biography. Instead of the remorselessly sequential narrative that typically characterizes biographical accounts, greater sensitivity to the *spaces of a life* could open up new and revealing ways of taking the *measure of a life.* Take Charles Darwin. Here the biographer encounters a number of different Darwins—Darwin the experimenter, Darwin the traveler, Darwin the invalid, Darwin the investor, Darwin the dupe of quack medicine. We find a "Beagle" Darwin and a "Down" Darwin, a "family" Darwin and a "scheming" Darwin, and, perhaps most significant of all, a "private" Darwin and a "public" Darwin. To different audiences Darwin presented himself in different guises. In different spaces different Darwins surface. A "life geography" of Darwin would thus have much to commend it. More generally, a greater awareness of the spaces of biography, of the places of identity, of the geography of selfhood, would enormously enrich our understanding of the mutual making of science and scientist.

A second way a geography of science might be further cultivated is by extending investigations into the geography of rationality. Such a development will revolve around what might be described as the "regionalizing of reason." In times past, rational thought was typically regarded as an enterprise that transcended particularity and remained untouched by local circumstance. This depiction is inadequate. Consider. When we try to understand people's behavior, it is essential that we take into account their motivations and intentions in acting the way they did. In trying to understand these, we need to make some sense of the settings that render such intentions intelligible both to the actor and to others. To put it another way, the reason a person gives for behaving in a certain way is setting dependent. This means that standards of practical rationality—what passes as a good reason for believing something—are spatially referenced. What a person is warranted in asserting will be hugely different depending on a multitude of contextual conditions. Accordingly, determining whether an individual acted rationally requires us to ascertain the prevailing standards at that particular time and place. If we are seeking to determine whether a certain belief is rational, the proper question must be whether it is a rational belief for a particular person, in a particular situation, at a particular time. A twelfth-century sailor, a fifteenth-century magician, a seventeenth-century clergyman, and a twentieth-century astronomer are warranted in believing different things about, say, the movements of the planets. Rationality is always *situated* rationality. And it is always embodied rationality. For too long philosophical portraits of knowledge were quadriplegic, dominated by the placeless image of a "brain in a vat." Such metaphors do little to catch the obligations we have in finding out about things—we need to engage in certain activities, move to certain sites, look at certain objects, talk with certain people.

All of this means that the idea that there is a single, unified scientific rationality is highly dubious. What has been promoted as scientific objectivity, as the "view from nowhere," turns out to have always been a "view from somewhere." The recognition that rationality is not disembodied but positioned has significant implications for understanding science and scientists. It means that the customary conventions of practical reasoning that scientists resort to in the different locations have to be taken more seriously. It also implies that different scientific traditions and practices, in different historical and geographical settings, deploy different understandings of evidence, demonstration, proof, objectivity, and so on. Scientific rationality cannot be conceived of independently of temporal and spatial location.

For many readers, I suspect, there is something disquieting about the direction we have been moving. Science has usually been seen as a detached enterprise, impartial and impersonal in its interrogation of nature. Any concern with spatial circumstance or local particularity seems philosophically compromising, inasmuch as it suggests that there is no such thing as scientific truth. It seems to imply that if scientific knowledge is a product of local circumstances, shaped by particular conditions, it can make no claims to veracity. This is not a necessary inference. It is entirely plausible to argue, for example, that what *passes* as knowledge, what a person is *warranted* in believing, what *counts* as good grounds for a claim are relative to the circumstances people find themselves in without insisting that truth itself is relative to such factors. There is an important distinction to be drawn between what is true and what one is warranted in asserting. At different points in time and in different geographical contexts, people have been justified in holding scientific opinions and beliefs that lack credibility in other space-time circumstances. A belief may be false without violating any of the standards for knowledge claims that a society or subculture has installed to ensure cognitive propriety. So while the philosopher is legitimately concerned with ultimate matters of truth, the geographer of science can justifiably focus on what is taken to be knowledge, on what is accorded the status of truth. Because what is judged to be rational has differed from time to time and place to place, we can uncover an historical geography of rationalizing practices. Because the witness of certain groups in society has been regarded as more dependable than that of others, we can delve into the social geography of testimony and trust. Because reliability was as much a moral as a scientific virtue, we can map the ways scientists sought to put in place disciplinary techniques to deliver credibility. And we can engage in all these endeavors without trespassing on truth in the conventional sense of the term.

* * *

Over the past three centuries and more, the image of science as a placeless activity has bitten deep in our culture. As an enterprise science seems to be the locus classicus of knowledge that is displaced, dislocated, disembedded. So much has this been the case that it goes against our intuition to think that science is marked by the particularities of location. Yet as this book has shown, the impact of place on science is inescapable. I have only begun to chart a few of its dimensions. I hope that others will continue to map the terrain. Of course I am not implying that everything about science can be reduced to matters of space, any more than I am saying that everything about reality can be expressed in map form. But I am convinced that attending to the spaces of science is of no small significance in coming to grips with the character of scientific endeavor and therefore of the modern world itself.

Bibliographical Essay

Chapter 1. A Geography of Science?

O. H. K. Spate's inaugural lecture, *The Compass of Geography* (Canberra: Australian National University, 1953), expressed the view that beyond the trivial, geography had little or no effect on the conduct of science. Émile Durkheim's classic insistence on the sociological immunity of natural science may be found in his *Selected Writings* (Cambridge: Cambridge University Press, 1972). The restriction of a sociology of science to explaining deviations from proper method or merely to science's response to shifting patterns of funding finds expression in Joseph Ben-David, *The Scientist's Role in Society: A Comparative Study* (Canberra: Prentice-Hall, 1971). For a robust philosophical defense of the "placelessness" of science see Thomas Nagel, *The View from Nowhere* (New York: Oxford University Press, 1986). In his account "The Essence and Soul of the Seventeenth-Century Scientific Revolution," *Science in Context* 1 (1987): 87–101, Zev Bechler similarly rejects local factors as of anything other than incidental significance. A different perspective is opened up in A. I. Sabra, "Situating Arabic Science: Locality versus Essence," *Isis* 87 (1996): 654–70.

The different ways Darwinism was construed in New Zealand and the American South can be appreciated from John Stenhouse, "Darwinism in

New Zealand, 1859–1900," in *Disseminating Darwinism: The Role of Place, Race, Religion, and Gender,* ed. Ronald L. Numbers and John Stenhouse (New York: Cambridge University Press, 1999), 61–89, and Lester D. Stephens, *Science, Race, and Religion in the American South: John Bachman and the Charleston Circle of Naturalists, 1815–1895* (Chapel Hill: University of North Carolina Press, 2000). Literature on different responses to Newton, Humboldt, and Darwin is surveyed later. The case of differential reaction to Einsteinian relativity is the subject of Andrew Warwick's two-part analysis, "Cambridge Mathematics and Cavendish Physics: Cunningham, Campbell and Einstein's Relativity, 1905–1911," *Studies in the History and Philosophy of Science* 23 (1992): 625–56 and 24 (1993): 1–25. Useful reflections on the role of place in field biology, including its significance in the work of Charles Elton and Raymond E. Lindeman, and on what he calls "practices of place" are available in Robert E. Kohler, "Place and Practice in Field Biology," *History of Science* 40 (2002): 189–210. The comment about cold fusion comes from this article.

SPACE MATTERS

The prevalence of spatial metaphors in the modern lexicon is abundantly illustrated in the essays drawn together in Rolland G. Paulston, ed., *Social Cartography: Mapping Ways of Seeing Social and Educational Change* (New York: Garland, 1996). Something of this metaphorical fecundity is particularly evident in the essays in this collection by David Turnbull, "Constructing Knowledge Spaces and Locating Sites of Resistance in the Modern Cartographic Transformation," Anne Sigismund Huff, "Ways of Mapping Strategic Thought," and Crystal Bartolovich, "Mapping the Spaces of Capital."

The importance of space in social interaction is the burden of a number of publications by Erving Goffman, especially *The Presentation of Self in Everyday Life* (London: Allen Lane, 1969) and *Relations in Public: Microstudies of the Public Order* (London: Allen Lane, 1971). In *The Interpretation of Cultures: Selected Essays* (New York: Basic Books, 1973), Clifford Geertz uses the idea of the "imaginative universe" to stress the importance of locality in making sense of human communication and its interpretation as a sign system. The giving of "particular sense to particular things in particular places" also animates Geertz's *Local Knowledge: Further Essays in Interpretive Anthropology* (New York: Basic Books, 1983). The role of space in constituting systems of human interaction is elucidated in Anthony Giddens, *The*

Constitution of Society: Outline of the Theory of Structuration (Oxford: Polity Press, 1984), and an important statement of such themes by a human geographer is Nigel Thrift's essay "On the Determination of Social Action in Space and Time," *Environment and Planning D: Society and Space* 1 (1983): 23–57. The idea of space as a social production motivates Henri Lefebvre's *The Production of Space* (Oxford: Blackwell, 1991) and Edward Soja's *Postmodern Geographies: The Reassertion of Space in Critical Social Theory* (London: Verso, 1989).

Useful introductions to more general theoretical conceptions of space include Robert D. Sack, *Conceptions of Space in Social Thought* (London: Macmillan, 1980); Derek Gregory and John Urry, eds., *Social Relations and Spatial Structures* (London: Macmillan, 1985); Edward Soja, "The Spatiality of Social Life: Towards a Transformative Retheorisation," in *Social Relations and Spatial Structures,* ed. Derek Gregory and John Urry (London: Macmillan, 1985), 90–122; Neil Smith, *Uneven Development: Nature, Capital and the Production of Space,* 2nd ed. (Oxford: Blackwell, 1990); and J. Nicholas Entrikin, *The Betweenness of Place: Towards a Geography of Modernity* (London: Macmillan, 1991). The concept of "communities of discourse" and the ways they "articulate" and "disarticulate" with their social environments is the subject of Robert Wuthnow's *Communities of Discourse: Ideology and Social Structure in the Reformation, the Enlightenment, and European Socialism* (Cambridge: Harvard University Press, 1989). While Wuthnow does not cast his analysis explicitly in spatial terms, he does observe that "attention must be given not only to the places where new ideologies were successfully institutionalized but also to those places where ideological innovations failed to take root" (6). The concepts of "enabling" and "constraining" are developed in Giddens's *Constitution of Society.*

The Seamus Heaney phrase appears in his essay "The Sense of Place," published in his *Preoccupations: Selected Prose, 1968–1978* (London: Faber and Faber, 1980). An analysis of some of the ways the global continuously impinges on the local is provided in David Harvey, *The Condition of Postmodernity: An Inquiry into the Origins of Social Change* (Oxford: Blackwell, 1989). The theme of imagined geographies is explored in various ways in Edward W. Said, *Culture and Imperialism* (London: Chatto and Windus, 1993), and Derek Gregory, *Geographical Imaginations* (Oxford: Blackwell, 1994). The particular case of Europe's encounter with the New World is the subject of several treatments, among them Stephen Greenblatt, *Marvellous Possessions: The Wonder of the New World* (Chicago: University of Chicago Press, 1991), Peter Mason, *Deconstructing America: Representations of the Other* (New York: Routledge, 1990), and Anthony Pagden, *European Encounters with the New World* (New Haven: Yale University Press, 1993). Cartographic dimensions of this episode are addressed in J. B. Harley, *Maps and the Columbian Encounter* (Milwaukee: Golda Meir Library, 1990). For the genesis of the idea of the Pacific, see the essays collected in Roy MacLeod and Philip F. Rehbock, eds., *Darwin's Laboratory: Evolutionary Theory and Natural History in the Pacific* (Honolulu: University of Hawaii Press, 1994), and for "darkest Africa," P. Brantlinger, "Victorians and Africans: The Genealogy of the Myth of the Dark Continent," *Critical Inquiry* 12 (1985): 166–203. The Western construction of the "Orient" is developed in Said's stimulating and controversial *Orientalism* (London: Routledge and Kegan Paul, 1978). For a critical perspective on the Orientalism "debate," see John M. Mackenzie, *Orientalism: History, Theory and the Arts* (Manchester: Manchester University Press, 1995). Something of the West's scientific crusade into Palestine is charted in Naomi Shepherd's *The Zealous Intruders: From Napoleon to the Dawn of Zionism—the Explorers, Archaeologists, Artists, Tourists, Pilgrims, and Visionaries Who Opened Palestine to the West* (New York: Harper and Row, 1987).

The connection between space, power, and knowledge is one of the dominant themes in the writings of Michel Foucault. His studies of the asylum and the prison present his analyses of these particular arenas: Michel Foucault, *The Birth of the Clinic: An Archaeology of Medical Perception* (London: Tavistock, 1973), and his *Discipline and Punish: The Birth of the Prison* (London: Penguin, 1991). More general statements of the conceptual links appear in his *Power/Knowledge: Selected Interviews and Other Writings, 1972–1977* (Brighton: Harvester Press, 1980) and "Of Other Spaces," *Diacritics* 16 (1986): 22–27 (translated by Jay Miskowiec). Geographical accounts that interact with such concerns include Felix Driver, "Geography's Empire: Histories of Geographical Knowledge," *Society and Space* 19 (1992): 23–40. The significance of the insight that theory travels is elucidated in Edward W. Said's "Travelling Theory," in *The World, the Text and the Critic* (London: Vintage, 1991).

GEOGRAPHIES OF SCIENCE

In my view an important moment in the conversation between geographers and historians of science was the address given by Steven Shapin at the annual conference of the Royal Geographical Society (with the Institute of British Geographers) in 1996 and published as "Placing the View from

Nowhere: Historical and Sociological Problems in the Location of Science," *Transactions of the Institute of British Geographers* 23 (1998): 1–8. Indications of geographers' increasing interest in the social history of science include the following: David N. Livingstone, "The Spaces of Knowledge: Contributions towards a Historical Geography of Science," *Society and Space* 13 (1995): 5–34; David Demeritt, "Social Theory and the Reconstruction of Science and Geography," *Transactions of the Institute of British Geographers,* n.s., 21 (1996): 484–503; Trevor J. Barnes, *Logics of Dislocation: Models, Metaphors, and Meanings of Economic Space* (New York: Guilford Press, 1996); Charles W. J. Withers, "Geography, Natural History and the Eighteenth-Century Enlightenment: Putting the World in Place," *History Workshop Journal* 39 (1995): 137–63; idem, "Notes toward a Historical Geography of Geography in Early Modern Scotland," *Scotlands* 3 (1996): 111–24; and idem, "Towards a History of Geography in the Public Sphere," *History of Science* 34 (1999): 45–78. Symbolic of the growing concern with the significance of space and place within science studies is the special issue of *Science in Context* in 1991 titled "The Place of Knowledge: The Spatial Setting and Its Relations to the Production of Knowledge" and edited by Adi Ophir and Steven Shapin. Also highly significant was the conference organized by the British Society for the History of Science in March 1994 on the theme "Making Space: Territorial Themes in the History of Science." The proceedings of this conference are gathered in Jon Agar and Crosbie Smith, eds., *Making Space for Science: Territorial Themes in the Shaping of Knowledge* (London: Macmillan, 1998). The geography of knowledge more generally is the subject of chapter 4 of Peter Burke's *A Social History of Knowledge* (Oxford: Polity, 2000), titled "Locating Knowledge: Centres and Peripheries." See also Thomas F. Gieryn, "Three Truth-Spots," *Journal of the History of the Behavioural Sciences* 38 (2002): 113-32.

Chapter 2. Site: Venues of Science

In his phenomenological psychology, Erwin Straus crucially discriminated between optical and acoustic spaces; see his essay "The Forms of Spatiality," which appears in his *Phenomenological Psychology: The Selected Papers of Erwin W. Straus* (London: Tavistock, 1966). Some of the philosophical connections between scientific meaning and the sites of scientific inquiry are explored in Nicholas Jardine, *The Scenes of Inquiry: On the Reality of Questions in the Sciences* (Oxford: Clarendon Press, 2000).

HOUSES OF EXPERIMENT

A useful general introduction to the significance of "the place of production" in scientific history is found in chapter 3 of Jan Golinski's *Making Natural Knowledge: Constructivism and the History of Science* (Cambridge: Cambridge University Press, 1998). The ideal of solitude in securing natural and spiritual knowledge is discussed in Steven Shapin, " 'The Mind Is Its Own Place': Science and Solitude in Seventeenth-Century England," *Science in Context* 4 (1990): 191–218. My account of John Dee's household relies on Deborah E. Harkness, "Managing an Experimental Household: The Dees of Mortlake and the Practice of Natural Philosophy," *Isis* 88 (1997): 247–62; she makes the telling observation that "when examining the household as a site of knowledge, we must not forget that the early modern household was, first and last, domestic and feminine space." Some stimulating observations on the evolution of household space are to be found in Yi-fu Tuan, *Segmented Worlds and Self: Group Life and Individual Consciousness* (Minneapolis: University of Minnesota Press, 1982). More standard histories of the Elizabethan household include Alice T. Friedman, *House and Household in Elizabethan England* (Chicago: University of Chicago Press, 1989). Owen Hannaway argues that the idea of sites specifically dedicated to experimentation has direct roots in alchemical practice in his essay "Laboratory Design and the Aim of Science: Andreas Libavius versus Tycho Brahe," *Isis* 77 (1986): 585–610. The creation of laboratory space during the Scientific Revolution is the subject of Steven Shapin's important essay "The House of Experiment in Seventeenth-Century England," *Isis* 79 (1988): 373–404; my account of Boyle's laboratory and the phrase "experimental public" are derived from this piece. More generally, something of the host of experimental sites in seventeenth-century England can be gleaned from Charles Webster, *The Great Instauration: Science, Medicine and Reform, 1626–1660* (London: Duckworth, 1975), and Michael Hunter, *Science and Society in Restoration England* (Cambridge: Cambridge University Press, 1981).

Michael Faraday's demonstrations are discussed by David Gooding in " 'In Nature's School': Faraday as an Experimentalist," in *Faraday Rediscovered: Essays on the Life and Work of Michael Faraday, 1797–1867,* ed. David Gooding and Frank A. L. James (London: Macmillan, 1985), and Iwan Rhys Morus, *Frankenstein's Children: Electricity, Exhibition, and Experiment in Early-Nineteenth Century London* (Princeton: Princeton University Press, 1998). The public performances of the nuclear industry are treated in H. M. Collins, "Public Experiments and Displays of Virtuosity: The Core-Set Revisited," *Social Studies of Science* 18 (1988): 725–48. Eighteenth-century resonances between scientific demonstrations and visual entertainment are, in part, the subject of Barbara Maria Stafford's *Artful Science: Enlightenment Entertainment and the Eclipse of Visual Education* (Cambridge: MIT Press, 1994). Also relevant is Iwan Morus, "Currents from the Underworld: Electricity and the Technology of Display in Early Victorian England," *Isis* 84 (1993): 50–69. The circumstances surrounding the genesis of the Cavendish Laboratory and William Thomson's lab in Glasgow are discussed in Simon Schaffer, "Physics Laboratories and the Victorian Country House," in *Making Space for Science: Territorial Themes in the Shaping of Knowledge,* ed. Crosbie Smith and Jon Agar (Basingstoke, U.K.: Macmillan Press, 1998), 149–180, and in Crosbie Smith, " 'Nowhere but in a Great Town': William Thomson's Spiral of Classroom Credibility," also in Smith and Agar, *Making Space for Science,* 118–46. Maxwell's metaphysical interests in the connections between algebra and geometry are noted by George Elder Davie in his apologia for the Scottish intellectual tradition, *The Democratic Intellect: Scotland and Her Universities in the Nineteenth Century* (Edinburgh: Edinburgh University Press, 1961). Other aspects of the creation of laboratory space for various purposes are treated in Graeme Gooday, "Teaching Telegraphy and Electrotechnics in the Physics Laboratory: William Ayrton and the Creation of an Academic Space for Electrical Engineering in Britain, 1873–1884," *History of Technology* 13 (1991): 73–111, and idem, "The Premisses of Premises: Spatial Issues in the Historical Construction of Laboratory Credibility," in Smith and Agar, *Making Space for Science,* 216–45.

The case that crucial significance is to be attached to the local crafts and skills that are mobilized in the laboratory is prosecuted by Joseph Rouse. See his *Knowledge and Power: Toward a Political Philosophy of Science* (Ithaca, N.Y.: Cornell University Press, 1987) and *Engaging Science: How to Understand Its Practices Philosophically* (Ithaca, N.Y.: Cornell University Press, 1996). More general connections between "trust" and the "disembedding" mechanisms of modernity feature in Anthony Giddens, *The Consequences of Modernity* (Oxford: Polity Press, 1990).

CABINETS OF ACCUMULATION

Key aspects of the early history of museums are treated in the work of Paula Findlen, who develops the idea that the "museum was located between silence and sound." I have particularly in mind her "The Museum: Its Classical Etymology and Renaissance Genealogy," *Journal of the History of Collections* 1 (1989): 59–78, and *Possessing Nature: Museums, Collecting, and Scientific Culture in Early Modern Italy* (Berkeley: University of California Press, 1994). Other relevant works include Oliver Impey and Arthur MacGregor, eds., *The Origins of Museums: The Cabinet of Curiosities in Sixteenth- and Seventeenth-Century Europe* (Oxford: Clarendon Press, 1985); Lorraine J. Daston, "Marvellous Facts and Miraculous Evidence in Early Modern Europe," *Critical Inquiry* 18 (1991): 93–124; and Sharon MacDonald, ed., *The Politics of Display: Museums, Science, Culture* (London: Routledge, 1998). Lewis Pyenson and Susan Sheets-Pyenson provide a useful overview of the museum in chapter 5 of their *Servants of Nature: A History of Scientific Institutions, Enterprises and Sensibilities* (New York: W. W. Norton, 1999), 125–49, and it is from this source that I have taken the idea of a window into the psyche. The exclusion of women from Renaissance museums is the subject of Paula Findlen, "Masculine Prerogatives: Gender, Space, and Knowledge in the Early Modern Museum," in *The Architecture of Science,* ed. Peter Galison and Emily Thompson (Cambridge: MIT Press, 1999), 29–57. Commenting on the role of wonder in the museum, Lorraine Daston tellingly observes in her article "The Factual Sensibility," *Isis* 79 (1988): 452–70: "Wide-eyed with wonder and open-mouthed with surprise, the admiring visitor paid the collector the sincerest compliment of speechlessness." More generally, "wonder" and "the wondrous" are the subjects of Lorraine Daston and Katharine Park, *Wonders and the Order of Nature, 1150–1750* (New York: Zone Books, 1998).

The idea of the British Empire as a "collective improvisation" and the role of an archiving mentality in its production are major themes in Thomas Richards, *The Imperial Archive: Knowledge and the Fantasy of Empire* (London: Verso, 1993). Peale's museum in Philadelphia is discussed in Charlotte M. Porter, *The Eagle's Nest: Natural History and American Ideas, 1812–1842* (University: University of Alabama Press, 1986), while Hyatt's plans for the reorganizing of the Boston Public Museum are referred to in Sally Gregory Kohlstedt, "Natural History Museums in the United States, 1850–1900," in *Scientific Colonialism: A Cross-Cultural Comparison,* ed. Nathan Reingold and Marc Rothenberg (Washington D.C.: Smithsonian Institution Press, 1987), 167–90. Connections between museum culture and the college curriculum in antebellum America are detailed in Sally Gregory Kohlstedt's "Curiosities and Cabinets: Natural History Museums and Education on the Antebellum Campus," *Isis* 79 (1988): 405–26. Circumstances in the American Museum of Natural History during the early decades of the twentieth century are dealt with in Ronald Rainger, *An Agenda for Antiquity: Henry Fairfield Osborn and Vertebrate Paleontology at the American Museum of Natural History, 1890–1935* (University: University of Alabama Press, 1991). Stephen T. Asma charts the history of natural history museums on a wider scale in his *Stuffed Animals and Pickled Heads: The Culture and Evolution of Natural History Museums* (New York: Oxford University Press, 2001).

The ways natural history museums constituted maps of geological knowledge are elucidated in Sophie Forgan, "Bricks and Bones: Architecture and Science in Victorian Britain," in Galison and Thompson, *The Architecture of Science,* 181–208. In that same collection (165–80) George W. Stocking Jr. discusses "The Spaces of Cultural Representation, circa 1887 and 1969: Reflections on Museum Arrangement and Anthropological Theory in the Boasian and Evolutionary Traditions." The Pitt-Rivers Museum is examined in David K. van Keuren, "Museums and Ideology: Augustus Pitt-Rivers, Anthropological Museums, and Social Change in Later Victorian Britain," *Victorian Studies* 28 (1984): 171–89. I have taken the Pitt-Rivers quotation about the educational role of the museum from this source. Also relevant is William Ryan Chapman, "Arranging Ethnology: A. H. L. F. Pitt-Rivers and the Typological Tradition," in *Objects and Others: Essays on Museums and Material Culture,* ed. George W. Stocking Jr. (Madison: University of Wisconsin Press, 1985), 5–48. Patrick Geddes's Outlook Tower is discussed in Helen Meller, *Patrick Geddes: Social Evolutionist and City Planner* (London: Routledge, 1990), and on this subject I have also learned much from Charles W. J. Withers, *Geography, Science and National Identity: Scotland since 1520* (Cambridge: Cambridge University Press, 2001). The significance of space in the Berlin Ethnological Museum is treated in Andrew Zimmerman, "Anthropology and the Place of Knowledge in Imperial Berlin" (Ph.D. diss., University of California, San Diego). See also his *Anthropology and Antihumanism in Imperial Germany* (Chicago: University of Chicago Press, 2001). Other useful studies include Susan Leigh Star and James R. Griesemer, "Institutional Ecology, 'Translations' and Boundary Objects: Amateurs and Professionals in Berkeley's Museum of Vertebrate Zoology, 1907–39," *Social Studies of Science* 19 (1989): 387–420, and Annie E. Coombes, *Reinventing Africa: Museums, Material Culture and Popular Imagination in Late Victorian and Edwardian England* (New Haven: Yale University Press, 1994).

The remarkable diversity of relations between architecture and science is clear in the wide-ranging set of essays in Galison and Thompson, *The Architecture of Science.* The architecture of scientific institutions has been the subject of investigation by Sophie Forgan in such works as "Context, Image and Function: A Preliminary Enquiry into the Architecture of Scientific Societies," *British Journal for the History of Science* 19 (1986): 89–113, and "The Architecture of Display: Museums, Universities and Objects in Nineteenth-Century Britain," *History of Science* 32 (1994): 139–62. On the natural history museum as a temple of science, see Susan Sheets-Pyenson, "Civilizing by Nature's Example: The Development of Colonial Museums of Natural History, 1850–1900," in Reingold and Rothenberg, *Scientific Colonialism,* 351–77; Carla Yanni, *Nature's Museum: Victorian Science and the Architecture of Display* (London: Athlone, 1999); and William T. Stearn, *The Natural History Museum at South Kensington: A History of the British Museum (Natural History), 1735–1980* (London: Heinemann, 1981). This theme also surfaces in Duncan F. Cameron, "The Museum: A Temple or a Forum," *Journal of World History* 14 (1972): 189–202, and Susan Sheets-Pyenson, "Cathedrals of Science: The Development of Colonial Natural History Museums during the Late Nineteenth Century," *History of Science* 25 (1987): 279–300. The strategies of the new scientific elite to secure the moral authority hitherto resident in the Victorian clergy have been scrutinized by a number of scholars. Among them are Frank Miller Turner, "The Victorian Conflict between Science and Religion: A Professional Dimension," *Isis* 69 (1978): 356–76, and T. W. Heyck, *The Transformation of Intellectual Life in Victorian England* (London: Croom Helm, 1982). Conceptual and ideological matters to do with museum representations of the past are considered in Stephen Bann's *The Inventions of History: Essays on the Representation of the Past* (Manchester: Manchester University Press, 1990). The idea of "object-based epistemology" is developed in Steven Conn, *Museums and American Intellectual Life, 1876–1926* (Chicago: University of Chicago Press, 1998); the quotations from Louis Agassiz and Edward Drinker Cope are taken from this source. In American anthropology the triumph of the university over the museum in the early decades of the twentieth century is noted in Curtis M. Hinsley, "The Museum Origins of Harvard Anthropology, 1866–1915," in *Science at Harvard University: Historical Perspectives,* ed. Clark A. Elliott and Margaret W. Rossiter (Bethlehem, Pa.: Lehigh University Press, 1992), 121–45.

FIELD OPERATIONS

Dorinda Outram provides a stimulating account of the contrast between the field naturalist and the sedentary naturalist in her essay "New Spaces in Natural History," in *Cultures of Natural History,* ed. N. Jardine, J. A. Secord and

E. C. Spary (Cambridge: Cambridge University Press, 1996), 249–65. It is from this piece that I have taken the extracts from Georges Cuvier and the contrast between "*passage over* terrain" and the "immobile *gaze.*" Henrika Kuklick and Robert E. Kohler have drawn together a very useful set of essays on the whole question of field science under the title *Science in the Field*, as vol. 11 of *Osiris,* 2nd ser., for 1996. Their introduction to the collection presents a stimulating set of reflections on the whole topic, including the observation about field scientists' taking their "domestic habits of mind" with them as they travel. The paper in that collection by Bruce Hevly, "The Heroic Science of Glacier Motion" (66–86), deals with the dispute among Forbes, Tyndall, and Hopkins. This debate also features in Frank Cunningham, *James David Forbes: Pioneer Scottish Glaciologist* (Edinburgh: Scottish Academic Press, 1990), and in Crosbie Smith, "William Hopkins and the Shaping of Dynamical Geology, 1830–1860," *British Journal for the History of Science* 22 (1989): 27–52. Resonances between the discipline of geography and the literature of travel adventure are reviewed in Richard Phillips, *Mapping Men and Empire: A Geography of Adventure* (London: Routledge, 1997). The role of women in the field is the theme of Marcia Myers Bonta, *Women in the Field: America's Pioneering Women Naturalists* (College Station: Texas A&M Press, 1991), and Jane Robinson, *Wayward Women* (Oxford: Oxford University Press, 1990). Cheryl McEwan has argued that the masculinist ethos of fieldwork has militated against women's participation in such sciences as geology and physical geography. See Cheryl McEwan, "Gender, Science and Physical Geography in Nineteenth-Century Britain," *Area* 30 (1998): 215–23. The idea of the field as a rite of passage is developed by Matthew Sparke, "Displacing the Field in Fieldwork: Masculinity, Metaphor and Space," in *Bodyspace,* ed. Nancy Duncan (London: Routledge, 1996), 212–33. The case of Mary Kingsley is discussed by Alison Blunt in *Travel, Gender, and Imperialism: Mary Kingsley and West Africa* (New York: Guildford Press, 1994), while the differences between the travel writings of Victorian women in Africa are the subject of Cheryl McEwan, "Encounters with West African Women: Textual Representations of Difference by White Women Abroad," in *Writing Women and Space: Colonial and Postcolonial Geographies,* ed. Alison Blunt and Gillian Rose (New York: Guilford Press, 1994), 73–100. The significance of the field club movement in early Victorian English science is highlighted in Colin A. Russell, *Science and Social Change: 1700–1900* (London: Macmillan, 1983), and David Elliston Allen, *The Naturalist in Britain: A Social History* (London: Allen Lane, 1976). More particularly on women's involvement in botanical field work see Ann B. Shteir,

Cultivating Women, Cultivating Science: Flora's Daughters and Botany in England, 1760–1860 (Baltimore: Johns Hopkins University Press, 1996). Relevant too, in a more general way, is Londa Schiebinger, *The Mind Has No Sex? Women in the Origins of Modern Science* (Cambridge: Harvard University Press, 1989). The importance of the relationships that Alfred Russel Wallace developed in his fieldwork is the subject of Jane Camerini, "Wallace in the Field," *Osiris* 11 (1996): 44–65. Other important recent contributions to understanding the field include Christopher R. Henke, "Making a Place for Science: The Field Trial," *Social Studies of Science* 30 (2000): 483–511; Richard W. Burkhardt Jr., " Ethology, Natural History, the Life Sciences, and the Problem of Place," *Journal of the History of Biology* 32 (1999): 489–508; Robert E. Kohler, *Landscapes and Labscapes: Exploring the Lab-Field Frontier in Biology* (Chicago: University of Chicago Press, 2002); and idem, "Labscapes: Naturalizing the Lab," *History of Science* 40 (2002): 473-501.

On links between tradition, practice, and rationality I have learned much from the writings of Hans-Georg Gadamer, Michael Polanyi, and Alasdair MacIntrye. In particular I have found the following valuable: Hans-Georg Gadamer, *Philosophical Hermeneutics,* trans. David E. Linge (Berkeley: University of California Press, 1977), and idem, *Reason in the Age of Science,* trans. Frederick G. Lawrence (Cambridge: MIT Press, 1981); Michael Polanyi, *Personal Knowledge: Towards a Post-critical Philosophy* (Chicago: University of Chicago Press, 1958); Alasdair MacIntyre, *Whose Justice? Which Rationality?* (Notre Dame, Ind.: University of Notre Dame Press, 1988).

The "constructedness" of the field, particularly in the social sciences, has been the subject of concern among feminists, who speak of the links between the politics of fieldwork and the politics of representation. Within geography, something of the character of the debate may be gleaned from the essays in the *Professional Geographer* for 1994, titled "Women in the Field: Critical Feminist Methodologies and Theoretical Perspectives," 54–102. The emergence of social field survey is considered in Martin Bulmer, Kevin Bales, and Kathryn Kish Sklar, *The Social Survey in Historical Perspective* (New York: Cambridge University Press, 1991). A useful set of reflections on the significance of fieldwork in anthropology is available in Akhil Gupta and James Ferguson, eds., *Anthropological Location: Boundaries and Grounds of a Field Science* (Berkeley: University of California Press, 1998). A. A. Roldán and H. F. Vermeulen, *Fieldwork and Footnotes: Studies in the History of European Anthropology* (London: Routledge, 1995), should also be consulted. Malinowski's role in establishing fieldwork as the anthropological method par excellence is highlighted in Henrika Kuklick, *The Savage Within: The Social History of British Anthropology, 1885–1945* (Cambridge: Cambridge University Press, 1991), and Joan Vincent, *Anthropology and Politics: Visions, Traditions, and Trends* (Tucson: University of Arizona Press, 1990). The significance of Rockefeller Funding in advancing Malinowski's vision is revealed in George W. Stocking Jr., *After Tylor: British Social Anthropology, 1888–1951* (Madison: University of Wisconsin Press, 1995). Valuable too is Stocking's essay "The Ethnographer's Magic: Fieldwork in British Anthropology from Tylor to Malinowski," in George W. Stocking Jr., ed., *Observers Observed: Essays on Ethnographic Fieldwork,* History of Anthropology, vol. 1 (Madison: University of Wisconsin Press, 1983),70–120, where he describes fieldwork as the "central ritual of the tribe." Kuklick discusses the suspicion of fieldwork by Victorian gentlemen-scholars in her chapter "After Ishmael: The Fieldwork Tradition and Its Future," in the collection edited by Gupta and Ferguson. The political motivation for certain types of radical fieldwork is manifest in the geographical writings of William Bunge, most especially *Fitzgerald: The Geography of a Revolution* (Cambridge, Mass.: Schenkman, 1971), and "The First Years of the Detroit Geographical Expedition: A Personal Report," in *Radical Geography: Alternative Viewpoints on Contemporary Social Issues,* ed. Richard Peet (1969; London: Methuen, 1978). Some of the problems of conducting social scientific fieldwork at "home" are discussed by Melissa R. Gilbert, "The Politics of Location: Doing Feminist Research at 'Home,'" *Professional Geographer* 46 (1994): 90–96. The idea of fieldwork as "spatial practice" is elucidated by James Clifford, who derives the concept from Michel de Certeau's *The Practice of Everyday Life* (Berkeley: University of California Press, 1984). Clifford's "Spatial Practices: Fieldwork, Travel, and the Disciplining of Anthropology" appears in Gupta and Ferguson, *Anthropological Locations.*

GARDENS OF DISPLAY

A useful brief introduction to the cultural history of the garden can be found in Andrew Cunningham's chapter "The Culture of Gardens," in *Cultures of Natural History,* ed. N. Jardine, J. A. Secord, and E. C. Spary (Cambridge: Cambridge University Press, 1996), 38–56. An excellent overview of the history and meaning of the botanical garden is provided in John Prest, *The Garden of Eden: The Botanic Garden and the Re-creation of Paradise* (New Haven: Yale University Press, 1981), who makes the comment about the garden's gathering into one place the scattered pieces of the jigsaw puzzle. A chronology of early botanical gardens can be found in *Hortus Botanicus: The Botanic Garden and the Book; Fifty Books from the Sterling Morton Library Exhibited at the Newberry Library for the Fiftieth Anniversary of the Morton Arboretum* (Lisle, Ill.: Morton Arboretum, 1972). Historical aspects of the use of the ark and the garden as biblical images of knowledge sites in the seventeenth century are developed in Jim Bennett and Scott Mandelbrote, *The Garden, the Ark, the Tower, the Temple: Biblical Metaphors of Knowledge in Early Modern Europe* (Oxford: Museum of the History of Science in association with the Bodleian Library, 1998). The role of the Tradescants is discussed in Prudence Leith-Ross, *The John Tradescants: Gardeners to the Rose and Lily Queen* (London: P. Owen, 1984). On the use of French formal gardens as maps of social standing, see Chandra Mukerji, "Reading and Writing with Nature: Social Claims and the French Formal Garden," *Theory and Society* 19 (1990): 651–79. The idea that plants were thought of as nations surfaces in Janet Browne, *The Secular Ark: Studies in the History of Biogeography* (New Haven: Yale University Press, 1983). On the history of particular botanical gardens see E. C. Spary's study of the Paris Jardin, *Utopia's Garden: French Natural History from Old Regime to Revolution* (Chicago: University of Chicago Press, 2000), which develops the idea of garden landscaping as delivering "simulacra" of different environmental conditions; Ray Desmond, *Kew: The History of the Royal Botanic Gardens* (London: Harvill Press, 1995), from which I have taken the quotation from Banks about the king at Kew and the Chinese emperor at Jehol; Lucile H. Brockway, *Science and Colonial Expansion: The Role of the British Royal Botanic Gardens* (New York: Academic Press, 1979); and Harold R. Fletcher and William H. Brown, *The Royal Botanic Garden Edinburgh, 1670–1970* (Edinburgh: Her Majesty's Stationery Office, 1970). Richard Drayton places the history of Kew in the wider context of British imperial politics and post-Enlightenment ideas of improvement in *Nature's Government: Science, Imperial Britain, and the "Improvement" of the World* (New Haven: Yale University Press, 2000). For the role of Kew Gardens in the "Banksian empire," see the essays in David Philip Miller and Peter Hanns Reill, eds., *Visions of Empire: Voyages, Botany, and Representations of Nature* (Cambridge: Cambridge University Press, 1996), especially (for the network of Banksian collectors) David Mackay, "Agents of Empire: The Banksian Collectors and Evaluation of New Lands," 38–57. The "satellites" of Kew are examined in Alan Frost's essay in this same volume, "The Antipodean Exchange: European Horticulture and Imperial Designs," 58–79. Kew's role as "the great exchange house of empire" comes through in Hector

Charles Cameron, *Sir Joseph Banks* (Sydney: Angus and Robertson, 1952). More generally on botanical gardens in the context of empire, see John Gascoigne's *Joseph Banks and the English Enlightenment: Useful Knowledge and Polite Culture* (Cambridge: Cambridge University Press, 1994), which stresses the Enlightenment thrust of Banks's various imperial designs, and the comprehensive survey by Donal P. McCracken, *Gardens of Empire: Botanical Institutions of the Victorian British Empire* (London: University of Leicester Press, 1997).

The standard general history of zoological gardens remains Gustave Loisel, *Histoire des ménageries de l'antiquité à nos jours,* 3 vols. (Paris: Octave Doin, 1912). A useful brief account, which also treats botanical gardens, is to be found in chapter 6 of Pyenson and Sheets-Pyenson, *Servants of Nature,* and in R. J. Hoage, Anne Roskell, and Jane Mansour, "Menageries and Zoos to 1900," in *New World, New Animals: From Menagerie to Zoological Park in the Nineteenth Century,* ed. R. J. Hoage and William A. Deiss (Baltimore: Johns Hopkins University Press, 1996). The essays in this collection also include several illuminating case studies: for example, the London Zoo is treated in Harriet Ritvo's chapter "The Order of Nature: Constructing the Collections of Victorian Zoos"; Michael A. Osborne deals with Paris in his chapter "Zoos in the Family: The Geoffroy Saint-Hilaire Clan and the Zoos of Paris"; and Linden Gillbank considers the Melbourne Zoo in "A Paradox of Purposes: Acclimatization Origins of the Melbourne Zoo." The best account of French acclimatization is Michael A. Osborne, *Nature, the Exotic, and the Science of French Colonialism* (Bloomington: Indiana University Press, 1994). Gillbank provides further detail on Australian dimensions of the subject in "The Origins of the Acclimatisation Society of Victoria: Practical Science in the Wake of the Gold Rush," *Historical Records of Australian Science* 6 (1986): 359–74, and "The Acclimatisation Society of Victoria," *Victoria Historical Journal* 51 (1980): 255–70. An overall survey of acclimatization societies is available in Christopher Lever, *They Dined on Eland: The Story of the Acclimatization Societies* (London: Quiller Press, 1992). I have examined some aspects of human acclimatization in "Human Acclimatization: Perspectives on a Contested Field of Inquiry in Science, Medicine and Geography," *History of Science* 25 (1987): 359–94, and "Tropical Climate and Moral Hygiene: The Anatomy of a Victorian Debate," *British Journal for the History of Science* 32 (1999): 93–110. Carl Hagenbeck's activities are discussed in Herman Reichenbach, "A Tale of Two Zoos: The Hamburg Zoological Garden and Carl Hagenbeck's Tierpark," and the Versailles zoo as attesting to the political power of Louis XIV appears in Thomas Veltre, "Menageries, Metaphors, and Meanings," both in the Hoage and Deiss collection. The case of Ota Benga is the subject of Phillips Verner Bradford and Harvey Blume, *Ota: The Pygmy in the Zoo* (New York: St. Martin's Press, 1992). Felix Driver has discussed a comparable case of the exhibiting of two African boys in the "Stanley and African Exhibition" held in London in 1890: see his *Geography Militant: Cultures of Exploration and Empire* (Oxford: Blackwell, 2001), chapter 7. Proposals for an ethnological exhibit by the Asiatic Society are discussed in Gyan Prakash, *Another Reason: Science and the Imagination of Modern India* (Princeton: Princeton University Press, 1999). Gregg Mitman explores the significance of the African Plains exhibit at the New York Zoological Society and the subsequent development of the wildlife park in "When Nature *Is* the Zoo: Vision and Power in the Art and Science of Natural History," *Osiris,* 2nd ser., 11 (1996): 117–43. The role of animals in the Victorian era is delightfully treated in Harriet Ritvo, *The Animal Estate: The English and Other Creatures in the Victorian Age* (Cambridge: Harvard University Press, 1987), and it is from this source that I have taken the comment about zoos celebrating "the imposition of human structure on the threatening chaos of nature."

SPACES OF DIAGNOSIS

The standard history of hospital architecture since ancient times remains John D. Thompson and Grace Goldin, *The Hospital: A Social and Architectural History* (New Haven: Yale University Press, 1975). A good range of recent scholarly work on the subject has been drawn together in Lindsay Granshaw and Roy Porter, eds., *The Hospital in History* (London: Routledge, 1989). Lindsay Granshaw's own survey, "The Hospital" in *Companion Encyclopedia of the History of Medicine,* ed. W. F. Bynum and Roy Porter (London: Routledge, 1993), 1173–95, is a useful review. "Medicine in the Hospital" is the subject of chapter 2 of W. F. Bynum, *Science and the Practice of Medicine in the Nineteenth Century* (Cambridge: Cambridge University Press, 1994) and is particularly strong on the French experience. On the American hospital system, major works include Charles E. Rosenberg, *The Care of Strangers: The Rise of America's Hospital System* (New York: Basic Books, 1987), which discusses the idea of patients as "moral minors," and Rosemary Stevens, *In Sickness and in Wealth: American Hospitals in the Twentieth Century* (New York: Basic Books, 1989). For Britain, see Britan Abel-Smith, *The Hospitals in England and Wales, 1800–1948* (Cambridge: Harvard University Press,

1964). A stimulating recent essay on hospital design as disclosing the cultural values of society and on hospitals as moral institutions is Allan M. Brandt and David C. Sloane, "Of Beds and Benches: Building the Modern American Hospital," in Galison and Thompson, *The Architecture of Science,* 281–305. Conceptually, many newer studies of hospitals as disciplinary regimes have found inspiration in the work of Michel Foucault, notably, *The Birth of the Clinic* and *Discipline and Punish.* The idea of the emergency room as an ethical space is developed in Michael Kelly and Ricardo Sanchez, "The Space of the Ethical Practice of Emergency Medicine," *Science in Context* 4 (1991): 79–100, and I have drawn the example of cardiac arrest from this essay. The "Waterloo" quotation from James Simpson is recorded in Roy Porter, *The Greatest Benefit to Mankind: A Medical History of Humanity from Antiquity to the Present* (London: HarperCollins, 1997), 369. This remarkable work contains a wealth of historical details about hospitals; the development of the asylum is treated on 494–510.

The diversity of medieval sites of madness is disclosed in Chris Philo, "The 'Chaotic Spaces' of Medieval Madness: Thoughts on the English and Welsh Experience," in Mikulás Teich, Roy Porter, and Bo Gustafsson, eds., *Nature and Society in Historical Context* (Cambridge: Cambridge University Press, 1997), 51–90. The "spectator sport" aspect of Bedlam is recorded in Edward G. O'Donoghue, *The Story of Bethlehem Hospital from Its Foundation in 1247* (London: Unwin, 1914). On Edinburgh's "spaces of reason and unreason" I have learned much from Chris Philo, "Edinburgh, Enlightenment, and the Geographies of Unreason," in *Geography and Enlightenment,* ed. David N. Livingstone and Charles W. J. Withers (Chicago: University of Chicago Press, 1999), 372–98. The asylum's internal spatial arrangements are treated in Chris Philo, " 'Enough to Drive One Mad': The Organisation of Space in Nineteenth-Century Lunatic Asylums," in *The Power of Geography: How Territory Shapes Social Life,* ed. Jennifer Wolch and Michael Dear (London: Unwin Hyman, 1989), 258–90. Medicomoral judgments about the appropriate environments in which asylums were to be located are the focus in Chris Philo, " 'Fit Localities for an Asylum': The Historical Geography of the Nineteenth-Century 'Mad-Business' in England as Viewed through the Pages of the *Asylum Journal,*" *Journal of Historical Geography* 13 (1987): 398–415, and in Hester Parr and Chris Philo, *"A Forbidding Fortress of Locks, Bars and Padded Cells": The Locational History of Mental Health Care in Nottingham,* Historical Geography Research Series, no. 32 ([Glasgow]: Institute of British Geographers, 1996).

THE BODY OF SCIENTIFIC KNOWLEDGE

The use of animals in scientific experimentation is surveyed in Hank Davis and Dianne Balfour, eds., *The Inevitable Bond: Examining Scientist-Animal Interactions* (New York: Cambridge University Press, 1992). The case of the fruit fly is the subject of Robert E. Kohler, *Lords of the Fly:* Drosophila *Genetics and the Experimental Life* (Chicago: University of Chicago Press, 1994). Julia Cream's essay "Women on Trial: A Private Pillory?" in *Mapping the Subject: Geographies of Cultural Transformation,* ed. Steve Pile and Nigel Thrift (London: Routledge, 1995), 158–69, deals with the case of contraceptive trials. Nazi medical experimentation is dealt with in Robert N. Proctor, *Racial Hygiene: Medicine under the Nazis* (Cambridge: Harvard University Press, 1988), and Benno Müller-Hill, *Murderous Science: Elimination by Scientific Selection of Jews, Gypsies, and Others; Germany, 1933–1945,* trans. George R. Fraser (Oxford: Oxford University Press, 1988). Issues of human experimentation more generally are the subject of M. H. Pappworth, *Human Guinea Pigs: Experimentation on Man* (London: Routledge and Kegan Paul, 1967).

On the idea that scientific instruments returned natural philosophers to a prelapsarian state, I have learned much from Simon Schaffer, "Regeneration: The Body of Natural Philosophers in Restoration England," in *Science Incarnate: Historical Embodiments of Natural Knowledge,* ed. Christopher Lawrence and Steven Shapin (Chicago: University of Chicago Press, 1998), 83–120. Alexander von Humboldt's experiments using his own body are recorded in Douglas Botting, *Humboldt and the Cosmos* (London: Sphere Books, 1973), 34, 101, 153–54. Dorinda Outram uses some of these stories to open up questions about the embodiment of scientific knowledge in her essay "On Being Perseus: New Knowledge, Dislocation, and Enlightenment Exploration," in Livingstone and Withers, *Geography and Enlightenment,* 281–94. Simon Schaffer considers the use of the body in electrical experiments and its socioepistemic implications in "Self Evidence," *Critical Inquiry* 18 (1992): 327–62. I have taken the quotation from Abbé Nollet from this article. Aspects of Nollet's electrical therapeutics are discussed in Patricia Fara, *An Entertainment for Angels: Electricity in the Enlightenment* (Cambridge: Icon Books, 2002). A useful introduction to the whole idea of the embodiment of knowledge is available in Steven Shapin and Christopher Lawrence's introductory essay "The Body of Knowledge" in *Science Incarnate: Historical Embodiments of Natural Knowledge,* ed. Christopher Lawrence and Steven Shapin (Chicago: University of Chicago Press, 1998),

1–19. Shapin's own contribution to this collection, "The Philosopher and the Chicken: On the Dietetics of Disembodied Knowledge" (21–50), provides an intriguing overview of the historical connections between asceticism and knowing.

The exclusion of women from science is the subject of Londa Schiebinger, *The Mind Has No Sex? Women in the Origins of Modern Science* (Cambridge: Harvard University Press, 1989), and *Nature's Body: Gender in the Making of Modern Science* (Boston: Beacon Press, 1993). The racial pronouncements of Hume, Kant, and Hegel are discussed in David N. Livingstone, "Race, Space and Moral Climatology: Notes toward a Genealogy," *Journal of Historical Geography* 28 (2002): 159–80. On the general theme of race and science, see the essays in Sandra Harding, ed., *The "Racial" Economy of Science* (Bloomington: Indiana University Press, 1993). The attitude of Victorian scientists toward women is the subject of Susan Sleeth Mosedale, "Science Corrupted: Victorian Biologists Consider 'the Woman Question,'" *Journal of the History of Biology* 11 (1978): 32–41; Evelleen Richards, "Darwin and the Descent of Woman," in *The Wider Domain of Evolutionary Thought,* ed. D. Oldroyd and J. Langham (Dordrecht: Reidel, 1983), 57–111; Evelleen Richards, "Huxley and Woman's Place in Science: The 'Woman Question' and the Control of Victorian Anthropology," in *History, Humanity and Evolution: Essays for John C. Greene,* ed. James R. Moore (Cambridge: Cambridge University Press, 1989), 253–84 (the Huxley quotations are found on 260 and 256); and in particular, Cynthia Eagle Russett, *Sexual Science: The Victorian Construction of Womanhood* (Cambridge: Harvard University Press, 1989). The quotations from Polanyi to the effect that instruments are extensions of bodily senses are found in Michael Polanyi, *The Study of Man: The Lindsay Memorial Lectures* (Chicago: University of Chicago Press, 1959), 31, 67.

OF OTHER SPACES

The use of churches as astronomical observatories is the subject of J. L. Heilbron, *The Sun in the Church: Cathedrals as Solar Observatories* (Cambridge: Harvard University Press, 1999). The ship as itself an instrument of scientific survey is analyzed in Richard Sorrenson, "The Ship as a Scientific Instrument in the Eighteenth Century," *Osiris,* 2nd ser., 11 (1996): 221–36. Cook's use of navigational calculations to produce coastlines is discussed in Paul Carter, *The Road to Botany Bay: An Essay in Spatial History* (London: Faber,

1987). The tent as a site of anthropological knowledge is treated in Lynette Schumaker, "A Tent with a View: Colonial Officers, Anthropologists, and the Making of the Field in Northern Rhodesia, 1937–1960," *Osiris,* 2nd ser., 11 (1996): 237–58. The royal court as a setting for scientific debate or the consumption of natural knowledge is dealt with in different ways in Mario Biagioli, *Galileo Courtier: The Practice of Science in the Culture of Absolutism* (Chicago: University of Chicago Press, 1993), and Charles W. J. Withers, "Geography, Royalty and Empire: Scotland and the Making of Great Britain, 1603–1661," *Scottish Geographical Magazine* 113 (1997): 22–32. Work on the coffeehouse as crucial to the emergence of the public sphere takes as its starting point Jürgen Habermas's *The Structural Transformation of the Public Sphere: An Inquiry into a Category of Bourgeois Society,* trans. Thomas Burger (Cambridge: MIT Press, 1989). The place of science in such public arenas is discussed in Roger Cooter and Stephen Pumphrey, "Separate Spheres and Public Places: Reflections on the History of Science Popularization and Science in Popular Culture," *History of Science* 32 (1994): 237–67, and Larry Stewart, "Public Lectures and Private Patronage in Newtonian England," *Isis* 75 (1986): 47–58. Steve Pincus provides a valuable commentary on various aspects of coffeehouse culture in " 'Coffee Politicians Does Create': Coffeehouses and Restoration Political Culture," *Journal of Modern History* 67 (1995): 807–34. George Steiner's observation about the coffeehouse appears in his interview with Richard Kearney in *States of Mind: Dialogues with Contemporary Thinkers on the European Mind* (Manchester: Manchester University Press, 1995), 83. On the public house as a scientific site, see Ann Secord, "Science in the Pub: Artisan Botanists in Early Nineteenth-Century Lancashire," *History of Science* 32 (1994): 269–315.

Chapter 3. Region: Cultures of Science

Something of the ways geographers now conceive of the idea of "region" may be gleaned from entries in the successive editions of *The Dictionary of Human Geography,* published by Blackwell. The idea of a regional psychology and genius loci was propounded in A. J. Herbertson, "Regional Environment, Heredity and Consciousness," *Geographical Teacher* 8 (1916): 147–53. Herbertson's outlook is discussed in David N. Livingstone, *The Geographical Tradition: Episodes in the History of a Contested Enterprise* (Oxford: Blackwell, 1992). Internationalism in science is the subject of Frank Greenaway, *Science International: A History of the International Council of*

Unions (New York: Cambridge University Press, 1996); Brigitte Schroeder-Gudehus, "Nationalism and Internationalism," in *Companion to the History of Modern Science*, ed. R. C. Olby et al. (London: Routledge, 1990), 898–908; and in idem, "International Science from the Franco-Prussian War to World War II: An Era of Organization," and Ronald E. Doel, "Internationalism After 1940," both in *The Cambridge History of Science,* vol. 8, *Modern Science in National and International Context*, ed. David N. Livingstone and Ronald L. Numbers (New York: Cambridge University Press), in press.

REGION, REVOLUTION, AND THE RISE OF SCIENTIFIC EUROPE

The significance of Chinese and Arabic science for developments in Europe is treated in various works including Joseph Needham, *Science and Civilisation in China* (Cambridge: Cambridge University Press, 1954–); Joseph Needham, *The Grand Titration: Science and Society in East and West* (London: Allen and Unwin, 1969); J. B. Harley and David Woodward, eds., *The History of Cartography,* vol. 2, bk. 1, *Cartography in the Traditional Islamic and South Asian Societies* (Chicago: University of Chicago Press, 1992); el-Bushra el-Said, "Perspectives on the Contributions of Arabs and Muslims to Geography," *Geojournal* 26 (1992): 157–66; and Scott L. Montgomery, *Science in Translation: Movements of Knowledge through Cultures and Time* (Chicago: University of Chicago Press, 2000). A brief useful survey is provided in David Goodman, "Europe's Awakening," in *The Rise of Scientific Europe, 1500–1800*, ed. David Goodman and Colin A. Russell (London: Hodder and Stoughton, 1991), 1–30. For a readable and lucid introduction to the "Scientific Revolution" more generally, see Steven Shapin, *The Scientific Revolution* (Chicago: University of Chicago Press, 1996), which contains a valuable bibliographical essay covering a variety of perspectives on the theme—traditional and revisionist. The vitality of recent debates on whether it is correct to speak of "the Scientific Revolution" may be gleaned from the following two collections: David C. Lindberg and Robert S. Westman, eds., *Reappraisals of the Scientific Revolution* (Cambridge: Cambridge University Press, 1990), and Margaret J. Osler, ed., *Rethinking the Scientific Revolution* (Cambridge: Cambridge University Press, 2000).

Helpful English-language surveys of Italian science in this period include David Goodman, "Crisis in Italy," in Goodman and Russell, *The Rise of Scientific Europe,* 91–116, and Giuliano Pancaldi, "Modern Science in Italy," in Livingstone and Numbers, *The Cambridge History of Science,* vol. 8.

The significance of Italian courtly patronage and anatomy theaters is prominent in Mario Biagioli, "Scientific Revolution, Social Bricolage, and Etiquette," in *The Scientific Revolution in National Context*, ed. Roy Porter and Milukás Teich (Cambridge: Cambridge University Press, 1992), 11–54. The anatomy theater is also the subject of Giovanna Ferrari, "Public Anatomy Lessons and the Carnival: The Anatomy Theater of Bologna," *Past and Present* 117 (1987): 50–117. The literature on Galileo is vast. *The Cambridge Companion to Galileo*, ed. Peter Machamer (Cambridge: Cambridge University Press, 1998), provides an excellent entry point to the literature and a very valuable bibliography. Two essays by David C. Goodman provide fine introductions to Iberian science in the period: "Iberian Science: Navigation, Empire and Counter-Reformation," in Goodman and Russell, *The Rise of Scientific Europe,* 117–44, and "The Scientific Revolution in Spain and Portugal," in Porter and Teich, *Scientific Revolution in National Context,* 158–77. More detailed is Goodman's *Power and Penury: Government, Technology and Science in Philip II's Spain* (Cambridge; Cambridge University Press, 1988). The significance of voyages of reconnaissance for Portuguese science is highlighted in R. Hooykaas, *Humanism and the Voyages of Discovery in 16th Century Portuguese Science and Letters* (Amsterdam: North Holland, 1979), and the triumph of navigators' firsthand experience over Scholastic authority in the birth of modern science features prominently in R. Hooykaas, "The Rise of Modern Science: When and Why?" *British Journal for the History of Science* 20 (1987): 453–73. Some rather extravagant claims in this connection are made in David W. Waters, "Science and the Techniques of Navigation," in *Art, Science, and History in the Renaissance*, ed. Charles S. Singleton (Baltimore: Johns Hopkins University Press, 1967), 189–237, and Daniel Banes, "The Portuguese Voyages of Discovery and the Emergence of Modern Science," *Journal of the Washington Academy of Sciences* 28 (1988): 47–58. Portuguese contributions to early tropical medicine are the subject of C. R. Boxer, *Two Pioneers of Tropical Medicine: Garcia d'Orta and Nicolas Monardes* (London: Wellcome Historical Medical Library, 1963). Short surveys of the Scientific Revolution in the English national context include Noel Coley, "Science in Seventeenth-Century England," in Goodman and Russell, *The Rise of Scientific Europe,* 197–226, and John Henry, "The Scientific Revolution in England," in Porter and Teich, *Scientific Revolution in National Context,* 178–209. Different perspectives on the role of religion in shaping English science in the period may be found in Charles Webster, *The Great Instauration: Science, Medicine and Reform, 1626–1660* (London: Duckworth, 1975); John Morgan, "Puritanism and Science: A Reinterpretation," *Histori-*

cal Journal 22 (1979): 535–60; Douglas S. Kemsley, "Religious Influences in the Rise of Modern Science: A Review and Criticism, Particularly of the 'Protestant-Puritan Ethic' Theory," *Annals of Science* 24 (1968): 199–226; Margaret C. Jacob and James R. Jacob, "The Anglican Origins of Modern Science: The Metaphysical Foundations of the Whig Constitution," *Isis* 71 (1980): 251–67; John Morgan, "The Puritan Thesis Revisited," in *Evangelicals and Science in Historical Perspective*, ed. David N. Livingstone, D. G. Hart and Mark A. Noll (New York: Oxford University Press, 1999), 43–74. The idea that the rise of natural science owed much to the new, literalist way the Protestant Reformers read scripture is advanced in Peter Harrison, *The Bible, Protestantism and the Rise of Natural Science* (Cambridge: Cambridge University Press, 1998). More generally, the best overview of the relations between science and religion is John Hedley Brooke, *Science and Religion: Some Historical Perspectives* (Cambridge: Cambridge University Press, 1991). The importance of gentlemanly codes of conduct for the practice of English science is stressed in Steven Shapin, *A Social History of Truth: Civility and Science in Seventeenth Century England* (Chicago: University of Chicago Press, 1994). For the idea of a geography of Enlightenment see William Clark, Jan Golinski, and Simon Schaffer, eds., *The Sciences in Enlightened Europe* (Chicago: University of Chicago Press, 1999).

POWER, POLITICS, AND PROVINCIAL SCIENCE

Among the many studies of English provincial science I have found the following particularly helpful: Arnold Thackray, "Natural Knowledge in Cultural Context: The Manchester Model," *American Historical Review* 79 (1974): 672–709, from which I have taken the quotations from Joseph Priestley; Steven Shapin, "The Pottery Philosophical Society, 1819–1835: An Examination of the Cultural Uses of Provincial Science," *Science Studies* 2 (1972): 311–36; Robert H. Kargon, *Science in Victorian Manchester: Enterprise and Expertise* (Manchester: Manchester University Press, 1977); Colin Russell, *Science and Social Change, 1700–1900* (London: Macmillan, 1983); and Ian Inkster and Jack Morrell, eds., *Metropolis and Province: Science in British Culture, 1780–1850,* (Philadelphia; University of Pennsylvania Press, 1983). The scientific culture of nineteenth-century Sheffield is the subject of several of the essays drawn together in Ian Inkster's *Scientific Culture and Urbanisation in Industrialising Britain,* Variorum Collected Studies Series (Aldershot, U.K.: Ashgate, 1997). Several of these writings refer to "the cultural geography of science." Other studies relevant to English provincial science include Jenny Uglow, *The Lunar Men: The Friends Who Made the Future* (London: Faber and Faber, 2002); Vladmir Jankovic, *Reading the Skies: A Cultural History of English Weather, 1650–1820* (Manchester: Manchester University Press, 2000); Simon Naylor, "The Field, the Museum and the Lecture Hall: The Spaces of Natural History in Victorian Cornwall," *Transactions of the Institute of British Geographers,* n.s., 27 (2002): 494-513. The ways the British Association for the Advancement of Science was conditioned by political and cultural affairs are foregrounded in Jack Morrell and Arnold Thackray, *Gentlemen of Science: Early Years of the British Association for the Advancement of Science* (Oxford: Clarendon Press, 1981). The political shape and what he calls "the social geography" of pre-Darwinian debates over evolution in Britain are the subject of Adrian Desmond's outstanding study, *The Politics of Evolution: Morphology, Medicine, and Reform in Radical London* (Chicago: University of Chicago Press, 1989). The social geography of London science in the early nineteenth century is laid out in Iwan Morus, Simon Schaffer, and James Secord, "Scientific London," in *London—World City, 1800–1840,* ed. C. Fox (New Haven: Yale University Press, 1992), 129–42. National styles of science have been treated by a number of authors, including Alistair Crombie, *Styles of Scientific Thinking in the European Tradition,* 3 vols. (London: Duckworth, 1994), and Nathan Reingold, "The Peculiarities of the Americans, or Are There National Styles in the Sciences?" *Science in Context* 4 (1991): 347–66. On the idea of cognitive styles more generally see Marga Vicedo, "Scientific Styles: Toward Some Common Ground in the History, Philosophy and Sociology of Science," *Perspectives on Science* 3 (1995): 231–54, and Ian Hacking, "Styles of Scientific Reasoning," in *Post-analytic Philosophy*, ed. John Rajchman and Cornel West (New York: Columbia University Press, 1985), 145–64. The relevance of scientific styles to particular disciplines can be found in Martin Rudwick, "Cognitive Styles in Geology," in *Essays in the Sociology of Perception*, ed. Mary Douglas (London: Routledge and Kegan Paul, 1982), 219–41; Mary Jo Nye, "National Styles? French and English Chemistry in the Nineteenth and Early Twentieth Centuries, *Osiris* 8 (1993): 30–52; Jonathan Harwood, *Styles of Scientific Thought: The German Genetics Community, 1900–1933* (Chicago: University of Chicago Press, 1993); and Malcolm Nicolson, "National Styles, Divergent Classifications: A Comparative Case Study from the History of French and American Plant Ecology," *Knowledge and Society: Studies in the Sociology of Science Past and Present* 8 (1989): 139–86. Some critical observations on the "national" scale of analysis in favor of "regional" and "local" scales can be found in Lewis Pyenson, "An End to National Science: The Meaning and the Extension of Local Knowledge," *History of Science* 40 (2002): 251-90. Here he argues that local urban conditions are often projected as national stereotypes.

REGION, READING, AND THE GEOGRAPHIES OF RECEPTION

The "intellectual geography" of Europe and its significance for scholarly migration are discussed in chapter 4 of Robert Mandrou's *From Humanism to Science: 1480–1700* (London: Penguin, 1978). The idea of a "geography of reception" and the particular case of Alexander von Humboldt are the subject of Nicolaas Rupke, "A Geography of Enlightenment: The Critical Reception of Alexander von Humboldt's Mexico Work," in Livingstone and Withers, *Geography and Enlightenment,* 319–39. The nature of Humboldtian science more generally is considered in M. Dettelbach, "Humboldtian science," in Jardine, Secord, and Spary, *Cultures of Natural History,* 287–304, and Susan Faye Cannon, *Science in Culture: The Early Victorian Period* (New York: Dawson and Science History Publications, 1978). The idea of "geographies of reading" is developed in the remarkable account of the reception of *Vestiges* by James A. Secord, *Victorian Sensation: The Extraordinary Publication, Reception, and Secret Authorship of "Vestiges of the Natural History of Creation"* (Chicago: University of Chicago Press, 2000). The book's European fortunes are the subject of Nicolaas Rupke, "Translation Studies in the History of Science: The Example of *Vestiges,*" *British Journal for the History of Science* 33 (2000): 209–22.

Studies of the ways Einsteinian and Darwinian theories have been received include Thomas F. Glick, *The Comparative Reception of Relativity,* Boston Studies in the Philosophy of Science (Dordrecht: Reidel, 1987); Thomas F. Glick, ed., *The Comparative Reception of Darwinism* (Chicago: University of Chicago Press, 1974); and Ronald L. Numbers and John Stenhouse, eds., *Disseminating Darwinism: The Role of Place, Race, Religion, and Gender* (New York: Cambridge University Press, 1999). I have discussed the responses to evolution by Calvinists in different cities in "Darwinism and Calvinism: The Belfast-Princeton Connection," *Isis* 83 (1992): 408–28, and "Science, Region, and Religion: The Reception of Darwinism in Princeton, Belfast, and Edinburgh," in Numbers and Stenhouse, *Disseminating Darwinism,* 7–38. Warfield's response to evolution is laid out in David N. Livingstone and Mark A. Noll, "B. B. Warfield (1851–1921): A Biblical Inerrantist as Evolutionist," *Isis* 91 (2000): 283–304. The racial fixations of natural historians in the southern states of America are detailed in Lester D. Stephens, *Science, Race, and Religion in the American South: John Bachman and the Charleston Circle of Naturalists, 1815–1895* (Chapel Hill: University of North Carolina Press, 2000). The reception of evolution in the American South is discussed in Ronald L. Numbers and Lester D. Stephens, "Darwinism in the American South," in Numbers and Stenhouse, *Disseminating Darwinism,* 123–43. The case of Alexander Winchell is treated in David N. Livingstone, *The Preadamite Theory and the Marriage of Science and Religion* (Philadelphia: American Philosophical Society, 1992), and Leonard Alberstadt, "Alexander Winchell's Preadamites—a Case for Dismissal from Vanderbilt University," *Earth Sciences History* 13 (1994): 97–112. New Zealand responses to evolution are dealt with in two articles by John Stenhouse: "The Darwinian Enlightenment and New Zealand Politics," in *Darwin's Laboratory: Evolutionary Theory and Natural History in the Pacific*, ed. Roy MacLeod and Philip F. Rehbock (Honolulu: University of Hawaii Press, 1994), and "Darwinism in New Zealand, 1859–1900," in Numbers and Stenhouse, *Disseminating Darwinism,* 61–89. The situation in Canada is discussed in Carl Berger, *Science, God, and Nature in Victorian Canada* (Toronto: University of Toronto Press, 1983); Michael Gauvreau, *The Evangelical Century: College and Creed in English Canada from the Great Revival to the Great Depression* (Montreal: McGill-Queen's University Press, 1991), chapter 4; and Suzanne Zeller, "Environment, Culture, and the Reception of Darwin in Canada, 1859–1909," in Numbers and Stenhouse, *Disseminating Darwinism,* 91–122. Russian reception is the subject of Alexander Vucinich, "Russia: Biological Sciences," in Glick, *Comparative Reception of Darwinism,* 227–55, and Daniel P. Todes, *Darwin without Malthus: The Struggle for Existence in Russian Evolutionary Thought* (Oxford: Oxford University Press, 1989).

SCIENCE, THE STATE, AND REGIONAL IDENTITY

A brief survey of national laboratories is available in Bob Seidel, "National Laboratories," in Livingstone and Numbers, *The Cambridge History of Science,* vol. 8. Conditions in France are treated in Roger Hahn, *The Paris Academy of Sciences* (Berkeley: University of California Press, 1968); for Germany, see David Cahan, *An Institute for an Empire: The Physikalisch-Technische Reichsanstalt, 1871–1918* (New York: Cambridge University Press, 1989). The scientific, political, and patriotic dimensions of national surveys for France, Scotland, and the United States come through in various ways in

Josef W. Konvitz, *Cartography in France, 1660–1848: Science, Engineering, and Statecraft* (Chicago: University of Chicago Press, 1987); J. Revel, "Knowledge of the Territory," *Science in Context* 4 (1991): 133–61; David Turnbull, "Cartography and Science in Early Modern Europe: Mapping the Construction of Knowledge Spaces," *Imago Mundi* 48 (1996): 5–24; Anne Marie Claire Godlewska, *Geography Unbound: French Geographic Science from Cassini to Humboldt* (Chicago: University of Chicago Press, 1999); Charles W. J. Withers, "How Scotland Came to Know Itself: Geography, National Identity and the Making of a Nation, 180–1790," *Journal of Historical Geography* 21 (1995): 371–97, and his *Geography, Science and National Identity: Scotland since 1520* (Cambridge: Cambridge University Press, 2001); John C. Greene, *American Science in the Age of Jefferson* (Ames: Iowa State University Press, 1984); Donald Jackson, *Thomas Jefferson and the Stony Mountains: Exploring the West from Monticello* (Urbana: University of Illinois Press, 1981); Silvio A. Bedini, *Thomas Jefferson. Statesman of Science* (New York: Macmillan, 1990); William H. Goetzmann, *Exploration and Empire: The Explorer and the Scientist in the Winning of the American West* (New York: Knopf, 1971); Hugh Richard Slotten, *Patronage, Practice, and the Culture of American Science: Alexander Dallas Bache and the U.S. Coast Survey* (New York: Cambridge University Press, 1994). Important, too, in this connection is David Buisseret, ed., *Monarchs, Ministers and Maps: The Emergence of Cartography as a Tool of Government in Early Modern Europe* (Chicago: University of Chicago Press, 1992).

The idea of the "scientific rationalization of society" was developed by Jürgen Habermas in the 1960s. See, for example, his essay "Technology and Science as 'Ideology,'" in *Toward a Rational Society: Student Protest, Science, and Politics,* trans. Jeremy J. Shapiro (Boston: Beacon Press, 1970), 81–122. The notion of governmentality is Michel Foucault's. A useful summary is to be found in his essay "Governmentality," trans. Rosi Braidotti, *Ideology and Consciousness* 3, no. 6 (1979): 5–21. German cameralism is discussed in Albion Small, *The Cameralists: The Pioneers of Social Polity* (Chicago: University of Chicago Press, 1909); Marc Raeff, *The Well-Ordered Police State: State and Institutional Change through Law in the Germanies and Russia, 1600–1800* (New Haven: Yale University Press, 1983); Richard Olson, *The Emergence of the Social Sciences, 1642–1792* (New York: Twayne, 1993), chapter 3, "Renaissance Naturalism and Political Economy in the German Cameralist Tradition"; and David F. Lindenfeld, *The Practical Imagination: The German Sciences of State in the Nineteenth Century* (Chicago: University of Chicago Press, 1997). A very helpful compact survey that pays considerable attention to the social uses of Germanic science is Kathryn M. Olesko, "Science in Germanic Europe," in Livingstone and Numbers, *The Cambridge History of Science,* vol. 8. The early history of political arithmetic is the subject of J. Mykkänen, " 'To Methodize and Regulate Them': William Petty's Governmental Science of Statistics," *History of the Human Sciences* 7 (1994): 65–88; Paul Buck, "Seventeenth-Century Political Arithmetic: Civil Strife and Vital Statistics," *Isis* 68 (1977): 67–84; Olson, *Emergence of the Social Sciences,* chapter 5, "Experimental Mechanical Philosophy, Political Arithmetic, and Political Economy in Seventeenth-Century Britain"; and Roger Smith, *The Fontana History of the Human Sciences* (London: Fontana Press, 1997), 307–14. On Linnaeus's economic thinking and his use of the idea of the polity of nature, I have learned much from Lisbet Koerner, *Linnaeus: Nature and Nation* (Cambridge: Harvard University Press, 1999). More generally on the theme of the "economy of nature," see Donald Worster, *Nature's Economy: A History of Ecological Ideas* (Cambridge, Cambridge University Press, 1977), and David N. Livingstone, "The Polity of Nature: Representation, Virtue, Strategy," *Ecumene* 2 (1995): 353–77. The cultural uses of science during the period of the Scientific Revolution are identified in Margaret C. Jacob, *The Cultural Meaning of the Scientific Revolution* (New York: Knopf, 1988); idem, *The Newtonians and the English Revolution, 1689–1720* (Ithaca, N.Y.: Cornell University Press, 1976); and Michael Hunter, *Science and Society in Restoration England* (Cambridge: Cambridge University Press, 1981). Steven Shapin provides a suggestive political reading of Newtonian debates in his essay "Of Gods and Kings: Natural Philosophy and Politics in the Leibniz-Clarke Disputes," *Isis* 72 (1981): 187–215. The promotion of science in the cause of cultural modernization projects in Argentina is revealed in Marcos Cueto, "Science in Spanish South America," forthcoming in Livingstone and Numbers, *The Cambridge History of Science,* vol. 8. The adoption of Lysenkoism by Soviet officialdom is treated in David Joravsky, *The Lysenko Affair* (Cambridge: Harvard University Press, 1970), and Loren Graham, *Science, Philosophy, and Human Behaviour in the Soviet Union* (Cambridge: Cambridge University Press, 1987).

Chapter 4. Circulation: Movements of Science

The story of the giraffe that Muhammad Ali presented to the king of France is delightfully told in Michael Allin, *Zarafa* (London: Headline, 1998), and the fate of the Fuegians who were brought to England and returned to

Tierra del Fuego in the early 1830s is the subject of Nick Hazelwood, *Savage: The Life and Times of Jemmy Button* (London: Hodder and Stoughton, 2000). Information about the early locations of Copernicus's *De Revolutionibus* comes from the census undertaken by Owen Gingerich: see his "The Great Copernican Chase," *American Scholar* 49 (1979–80): 81–88, and "The Censorship of Copernicus's *De Revolutionibus*," in *The Eye of Heaven: Ptolemy, Copernicus, Kepler* (New York: American Institute of Physics, 1993), 269–85. A brief summary of the European dissemination of Copernicanism can be found in Colin A. Russell, "The Spread of Copernicanism in Northern Europe," in *The Rise of Scientific Europe,* 63–90. The diffusion of the air pump is taken up in chapter 6 of Steven Shapin and Simon Schaffer, *Leviathan and the Air-Pump: Hobbes, Boyle, and the Experimental Life* (Princeton: Princeton University Press, 1985). The central importance of the air pump to the new experimental philosophy of the Scientific Revolution has frequently been remarked on, as, for example, in Rupert Hall, *From Galileo to Newton, 1630–1720* (London: Collins, 1963) and idem, *The Revolution in Science, 1500–1750* (London: Longman, 1983). More generally, the significance of the diffusion of science is treated, in different ways, in Ian Inkster, "Mental Capital: Transfers of Knowledge and Technique in Eighteenth-Century Europe," *Journal of European Economic History* 19 (1990): 403–41; Richard D. Brown, *Knowledge Is Power: The Diffusion of Information in Early America, 1700–1865* (New York: Oxford University Press, 1989); Raymond James Evans, "The Diffusion of Science: The Geographical Transmission of Natural Philosophy into the English Provinces, 1660–1760," Ph.D. diss., University of Cambridge (I am most grateful to Professor Andrew Cliff for making this work available to me).

TRANSLOCATION AND TRANSFERENCE: THE PROBLEM STATED

On the conceptual problems posed by the circulation of laboratory knowledge, see Shapin and Schaffer, *Leviathan and the Air-Pump.* The claim that the universalism of science is to do with "the adaptation of one local knowledge to create another" is made in Rouse, *Knowledge and Power,* 72. The significance of traveling during the period of the Scientific Revolution has only recently begun to be examined with the care it deserves. Traditionally, historians of the Scientific Revolution have tended to focus their inquiries on the experimental sciences. A few recent correctives include Anthony Grafton, *New Worlds, Ancient Texts: The Power of Tradition and the Shock of Discovery*

(Cambridge, Mass.: Belknap Press, 1992); Pyenson and Sheets-Pyenson, *Servants of Nature,* chapter 9, "Travelling: Discovery, Maps and Scientific Exploration"; and Lisa Jardine, *Ingenious Pursuits: Building the Scientific Revolution* (London: Little, Brown, 1999), chapters 5–7. Francis Bacon's comment on the significance of "distant voyages" occurs in aphorism 84 of *The New Organon.* The skeptical consequences of travel and the way voyages tended to bring transcendental concepts down to earth was highlighted by Paul Hazard in *The European Mind: 1680–1715* (Cleveland: Meridian Books, 1969), which originally appeared in 1935 as *La crise de la conscience européenne.* The compositional background to Cook's narratives is treated in Daniel Clayton, *Islands of Truth* (Vancouver: University of British Columbia Press, 1999). More generally on the issue of travel and knowledge, see I. S. MacClaren, "Exploration/Travel Literature and the Evolution of the Author," *International Journal of Canadian Studies* 5 (1992): 39–68; the essays in James Duncan and Derek Gregory, eds., *Writes of Passage: Reading Travel Writing* (London: Routledge, 1999); and Jas Elsner and Joan-Pau Rubiés, eds., *Voyages and Visions: Towards a Cultural History of Travel* (London: Reaktion Books, 1999). For a stimulating account of the contradictions between the *rhetoric* of rejecting trust, tradition, and authority and the *historical* reliance on precisely these things in scientific inquiry, see Shapin's *Social History of Truth.*

TRAVEL AND THE TECHNIQUES OF TRUST

Disciplining the Senses

Generally, on the issue of reports from travelers, consult R. W. Frantz, *The English Traveller and the Movement of Ideas, 1660–1732* (Lincoln: University of Nebraska Press, 1934); Percy G. Adams, *Travellers and Travel Liars, 1660–1800* (Berkeley: University of California Press, 1962); and Neil Rennie, *Far-Fetched Facts: The Literature of Travel and the Idea of the South Seas* (Oxford: Oxford University Press, 1995). Some of the ways observers were disciplined or drilled are discussed in Justin Stagl, "The Methodising of Travel in the Sixteenth Century," *History and Anthropology* 4 (1990): 303–38; idem, *A History of Curiosity: The Theory of Travel, 1550–1800,* Studies in Anthropology and History 13 (London: Routledge, 1995); Joan-Pau Rubiés, "Instructions for Travellers: Teaching the Eye to See," *History and Anthropology* 9 (1996): 139–90; Steven J. Harris, "Long-Distance Corporations, Big Sciences, and the Geography of Knowledge," *Configurations* 6 (1998): 269–304; D. Carey, "Compiling Nature's History: Travellers and Travel Narratives in the Early Royal Society," *Annals of Science* 54 (1997): 269–92; David S. Lux and Harold J. Cook, "Closed Circles or Open Networks? Communicating at a Distance during the Scientific Revolution," *History of Science* 36 (1998): 179–211. John Law's analysis centers on the crucial significance of "documents, devices and drilled people": see his "On the Methods of Long-Distance Control: Vessels, Navigation and the Portuguese Route to India," in *Power, Action and Belief: A New Sociology of Knowledge?* ed. John Law, Sociological Review Monograph 32 (London: Routledge and Kegan Paul, 1986), 234–63. The role of the Jesuits as distant observers is treated in Steven J. Harris, "Confession-Building, Long-Distance Networks, and the Organization of Jesuit Science," *Early Science and Medicine* 1 (1996): 299–304, and Alice Stroup, *A Company of Scientists: Botany, Patronage, and Community at the Seventeenth-Century Parisian Royal Academy of Sciences* (Berkeley: University of California Press, 1990). Thomas Pennant's Scottish survey is examined in Charles W. J. Withers, "Travel and Trust in the Eighteenth Century," in *L'Invitation au Voyage: Studies in Honour of Peter France,* ed. John Renwick (Oxford: Voltaire Foundation, 2000), 47–54. Other examples of the use of circulated queries in the Scottish context may be found in Withers, *Geography, Science and National Identity.* For the nineteenth century and the use of *Hints to Travellers,* I have benefited from Driver, *Geography Militant,* chapter 3. See also Jonathan Crary, *Techniques of the Observer: On Vision and Modernity in the Nineteenth Century* (Cambridge: MIT Press, 1992). The controversy over Timbuctoo and the role of wounds in establishing trust is the subject of Michael J. Heffernan's arresting paper, " 'A Dream as Frail as Those of Ancient Time': The In-credible Geographies of Timbuctoo," *Environment and Planning D: Society and Space* 19 (2001): 203–25. See also Gerd Spittler, "Explorers in Transit: Travels to Timbucktu and Agades in the Nineteenth Century," *History and Anthropology* 9 (1996): 231–53.

Mapping Territory

Standard treatments of the history of cartography include Leo Bagrow, *History of Cartography* (Cambridge: Harvard University Press, 1964); P. D. A. Harvey, *The History of Topographical Maps: Symbols, Pictures and Surveys* (London: Thames and Hudson, 1980); John Noble Wilford, *The Mapmakers* (New York: Knopf, 1981); Norman J. W. Thrower, *Maps and Civilization:*

Cartography in Culture and Society (Chicago: University of Chicago Press, 1996). The most exciting general treatment is the multivolume *History of Cartography* currently being produced by the University of Chicago Press. Mercator's map projections are discussed in the biography by Nicholas Crane, *Mercator: The Man Who Mapped the Planet* (London: Weidenfeld and Nicolson, 2002). Reassessments of the conventional understanding of maps as scientific documents owe much to the work of the late J. B. Harley. Among his most important contributions are the following articles: "Silences and Secrecy: The Hidden Agenda of Cartography in Early Modern Europe," *Imago Mundi* 40 (1988): 57–76; "Maps, Knowledge and Power," in *The Iconography of Landscape,* ed. Denis Cosgrove and Stephen Daniels (Cambridge: Cambridge University Press, 1988), 277–312; "Deconstructing the Map," *Cartographica* 26 (1989): 1–20; and "Cartography, Ethics and Social Theory," *Cartographica* 27 (1990): 1–23. Other significant statements are available in Denis Wood, *The Power of Maps* (London: Routledge, 1992); Matthew H. Edney, "Cartography without 'Progress': Reinterpreting the Nature and Historical Development of Mapmaking," *Cartographica* 30 (1993): 54–68; Simon Berthon and Andrew Robinson, *The Shape of the World: The Mapping and Discovery of the Earth* (London: George Philip, 1991); Chandra Mukerji, "Visual Language in Science and the Exercise of Power: The Case of Cartography in Early Modern Europe," *Studies in Visual Communication* 10 (1984): 30–45; David Turnbull, *Maps Are Territories: Science Is an Atlas* (Chicago: University of Chicago Press, 1989); and Denis Cosgrove, ed., *Mappings* (London: Reaktion Books, 1999). Two outstanding studies of the constructive capacities of early modern cartography are Frank Lestringant, *Mapping the Renaissance World: The Geographical Imagination in the Age of Discovery* (Oxford: Polity Press, 1994), and Jerry Brotton, *Trading Territories: Mapping the Early Modern World* (London: Reaktion Books, 1997). The Lewis Carroll quotation is from *Sylvie and Bruno* and may be found in *The Complete Works of Lewis Carroll* (New York: Random House, 1939), 7:556–57.

The early mapping of the Americas is the subject of J. B. Harley, *Maps and the Columbian Encounter* (Milwaukee: Golda Meir Library, 1990). James Cook's place-naming activities in Australia and New Zealand are reviewed in Paul Carter, *The Road to Botany Bay: An Essay in Spatial History* (London: Faber, 1987). The standard account of the mapping of India is now Matthew Edney, *Mapping an Empire: The Geographical Construction of British India, 1765–1843* (Chicago: University of Chicago Press, 1997). George Vancouver's mapping enterprises have been discussed in Daniel Clayton, "On the

Colonial Genealogy of George Vancouver's Chart of the North-West Coast of North America," *Ecumene* 7 (2000): 371–401. The conventional character of geological mapping is the subject of Martin Rudwick, "The Emergence of a Visual Language for Geological Science," *History of Science* 14 (1976): 149–95. The case of La Pérouse in the Pacific is treated in Bruno Latour, "Visualisation and Cognition: Thinking with Eyes and Hands," in *Knowledge and Society: Studies in the Sociology of Culture Past and Present,* vol. 6, ed. H. Kuklick and E. Long (Greenwich, Conn.: JAI Press, 1986), and in Bruno Latour, *Science in Action: How to Follow Scientists and Engineers through Society* (Cambridge: Harvard University Press, 1987). Michael T. Bravo stresses the linguistic and ethnographic dimensions of La Pérouse's experience on Sakhalin in "Ethnographic Navigation and the Geographical Gift," in Livingstone and Withers, *Geography and Enlightenment,* 199–235. Schomburgk's surveying of British Guiana is superbly told in D. Graham Burnett, *Masters of All They Surveyed: Exploration, Geography, and a British El Dorado* (Chicago: University of Chicago Press, 2000), and the Thailand case is the subject of Thongchai Winichakul, *Siam Mapped: A History of the Geo-body of a Nation* (Honolulu: University of Hawaii Press, 1994).

The quotations from Polanyi and Kuhn on the resonances between maps and scientific theories come from Michael Polanyi, *Personal Knowledge: Towards a Post-critical Philosophy* (London: Routledge and Kegan Paul, 1958), 4; and Thomas S. Kuhn, *The Structure of Scientific Revolutions,* 2nd ed. (Chicago: University of Chicago Press, 1970), 109. Humboldt's use of the isoline is discussed in Godlewska, *Geography Unbound,* 254–55, and his use of the technique to create an "isoworld" is tellingly expounded in Michael Dettelbach, "Global Physics and Aesthetic Empire: Humboldt's Physical Portrait of the Tropics," in Miller and Reill, *Visions of Empire,* 258–92. On the "Wallace line" and its genealogy, see Jane R. Camerini, "Evolution, Biogeography, and Maps: An Early History of Wallace's Line," *Isis* 84 (1993): 700–727, and James Moore, "Wallace's Malthusian Moment: The Common Context Revisited," in *Victorian Science in Context,* ed. Bernard Lightman (Chicago: University of Chicago Press, 1997), 290–311. The Wallace quotation about every species coming into existence coincident in both space and time comes from Alfred R. Wallace, "On the Law Which Has Regulated the Introduction of New Species," *Annals and Magazine of Natural History,* 2nd ser., 16 (1855): 184–96, on 186. The imperial science of Roderick Murchison is the subject of James Secord, "King of Siluria: Roderick Murchison and the Imperial Theme in Nineteenth-Century British Geology," *Victorian Studies* 25 (1982): 413–42, and Robert A. Stafford, *Scientist of Empire: Sir Roderick*

Murchison, Scientific Exploration and Victorian Imperialism (Cambridge: Cambridge University Press, 1989). The general issue of dividing the world up in various ways during the imperial era animates John Willinsky's *Learning to Divide the World: Education at Empire's End* (Minneapolis: University of Minnesota Press, 1998).

Picturing the Unfamiliar

My quotations from the *Art Journal* for 1856 and 1860 are taken from Joan M. Schwartz, "*The Geography Lesson:* Photographs and the Construction of Imaginative Geographies," *Journal of Historical Geography* 22 (1996): 16–45. The phrase about travel photographs' reducing "sites to sights" also comes from this piece. Major works on the scientific illustrations of travelers include Bernard Smith, *European Vision and the South Pacific,* 2nd ed. (New Haven: Yale University Press, 1985); Barbara Maria Stafford, *Voyage into Substance. Art, Science, Nature, and the Illustrated Travel Account, 1760–1840* (Cambridge: MIT Press, 1984); James Krasner, *The Entangled Eye: Visual Perception and the Representation of Nature in Post-Darwinian Narrative* (Oxford: Oxford University Press, 1992); and Katherine Manthorne, *Tropical Renaissance: North American Artists Exploring Latin America, 1839–1879* (Washington, D.C.: Smithsonian Institution Press, 1989). The trust relationship in natural history illustration is highlighted in Martin Kemp, " 'Taking It on Trust': Form and Meaning in Naturalistic Representation," *Archives of Natural History* 17 (1990): 127–88. Some suggestive observations on botanical illustration are also to be found in Martin Kemp, " 'Implanted in Our Natures': Humans, Plants, and the Stories of Art," and Simon Schaffer, "Visions of Empire: Afterword," both in Miller and Reill, *Visions of Empire.* The quotation from Hawkesworth occurs in the first volume of *An Account of the Voyages Undertaken by the Order of His Present Majesty for Making Discoveries in the South Hemisphere* (London, 1773), xvi.

The use of photographic representation in scientific inquiry has been probed by a number of authors including Lorraine Daston and Peter Galison, "The Image of Objectivity," *Representations* 40 (1992): 81–128, and Jonathan Crary, *Techniques of the Observer: Vision and Modernity in the Nineteenth Century* (Ithaca, N.Y.: Cornell University Press, 1991). Scientific discoveries using photography are treated in Jon Darius, *Beyond Vision* (Oxford: Oxford University Press, 1984). Specifically on its use in meteorology, see Jennifer Tucker, "Photography as Witness, Detective, and Impostor: Visual

Representation in Victorian Science," in Lightman, *Victorian Science,* 378–408; I have taken the quotation about inserting a yard measure from this source. For anthropology, see Elizabeth Edwards, ed., *Anthropology and Photography, 1860–1920* (New Haven: Yale University Press, 1992); for astronomy, John Lankford, "Photography and the Nineteenth-Century Transits of Venus," *Technology and Culture* 28 (1987): 648–57, and Alex Soojung-Kim Pang, "Victorian Observing Practices, Printing Technology and Representations of the Solar Corona," *Journal of the History of Astronomy* 25 (1994): 249–74; for medicine, Lisa Cartwright, *Screening the Body: Tracing Medicine's Visual Culture* (Minneapolis: University of Minnesota Press, 1995); for geography and imperialism, James R. Ryan, *Picturing Empire: Photography and the Visualization of the British Empire* (London: Reaktion Books, 1997). The strategies of the *National Geographic* are the subject of Catherine A. Lutz and Jane L. Collins, *Reading "National Geographic"* (Chicago: University of Chicago Press, 1993), and S. Montgomery, "Through a Lens, Brightly: The World according to *National Geographic,*" *Science as Culture* 4 (1993): 4–46.

GATHERING THE WORLD TOGETHER

The idea of "centers of calculation" is developed in Latour, *Science in Action.* My treatment of the Casa de la Contratación as an early "knowledge space" draws on David Turnbull, "Cartography and Science in Early Modern Europe: Mapping the Construction of Knowledge Spaces," *Imago Mundi* 46 (1996): 5–24. Portolan charts are the subject of Tony Campbell, "Portolan Charts from the Late Thirteenth Century to 1500," in Harley and Woodward, *The History of Cartography,* vol. 1. The imperial character of Kew in general and of Banks in particular comes through clearly in Desmond, *Kew.* Both the observation about Banks as "the staunchest imperialist of the day" and the comments from the Foreign Office about the colonial value of botanical knowledge are taken from this source. On Banks's home as a "center of calculation," see David Philip Miller, "Joseph Banks, Empire, and 'Centres of Calculation' in Late Hannoverian London," in Miller and Reill, *Visions of Empire,* 21–37. The cultural and scientific significance of debates over the imperial yard are detailed in Simon Schaffer, "Metrology, Metrication and Values," in Lightman, *Victorian Science,* 438–74. Relevant too are Allan Megill, ed., *Rethinking Objectivity* (Durham, N.C.: Duke University Press, 1994), and Julian Hoppit, "Reforming Britain's Weights and Measures," *English Historical Review* 108 (1993): 82–104. Ken Alder tells the story of the determination of the meter in *The Measure of All Things: The Seven-Year Odyssey That Transformed the World* (London: Little, Brown, 2002). The ideal of precision in the reports of late Enlightenment scientific travelers is the subject of Michael T. Bravo, "Precision and Curiosity in Scientific Travel: James Rennell and the Orientalist Geography of the New Imperial Age (1760–1830)," in Elsner and Rubiés, *Voyages and Visions,* 162–83. Connections between metrology and circulation are examined in Joseph O'Connell, "Metrology: The Creation of Universality by the Circulation of Particulars," *Social Studies of Science* 23 (1993): 129–73. The use of the Munsell code as a means of "circulating reference" among field scientists in Amazonia is the subject of a personal narrative by Bruno Latour, "Circulating Reference: Sampling the Soil in the Amazon Forest," in his *Pandora's Hope: Essays on the Reality of Science Studies* (Cambridge: Harvard University Press, 1999), chapter 2. Useful brief surveys of measurement are to be found in chapter 7 of Pyenson and Sheets-Pyenson, *Servants of Nature* and for earlier periods, in Alfred W. Crosby, *The Measure of Reality: Quantification and Western Society, 1250–1600* (Cambridge: Cambridge University Press, 1997). The cultural history of the drive for quantitative rigor in both society and science more generally is tellingly elucidated in Theodore M. Porter, *Trust in Numbers: The Pursuit of Objectivity in Science and Public Life* (Princeton: Princeton University Press, 1995).

Chapter 5. Putting Science in Its Place

The concept of "immutable mobiles" is developed in Latour, *Science in Action.* Philosophical interest in questions of place is particularly conspicuous in Edward S. Casey's *Getting Back into Place: Toward a Renewed Understanding of the Place-World* (Bloomington: Indiana University Press, 1993), which subjects "the place we occupy" to philosophical scrutiny because, Casey insists, it "has everything to do with what and who we are." The prevalence of spatial expressions in efforts to elucidate the experience of modernity—including the phrases "the geography of social statuses and functions" and "the space of moral and spiritual orientation"—is evident in Charles Taylor, *Sources of the Self: The Making of Modern Identity* (Cambridge: Harvard University Press, 1989). The mapping of the London locations of the key figures in the Devonian controversy was undertaken by Martin J. S. Rudwick, *The Great Devonian Controversy: The Shaping of Scientific Knowledge among Gentlemanly Specialists* (Chicago: University of Chicago Press, 1985). Here Rudwick develops the idea of the "cognitive topography" of geological expertise. Something of the different spaces of Darwin's persona can be gleaned from Adrian Desmond and James Moore, *Darwin* (London: Michael Joseph, 1991); Janet Browne, *Charles Darwin,* vol. 1, *Voyaging* (New York: Knopf, 1995); and *Charles Darwin,* vol. 2, *The Power of Place* (New York: Knopf, 2002). The importance of "settings" in making sense of rational inquiry is crucial in the writings of Alasdair MacIntyre. See his *After Virtue: A Study in Moral Theory,* 2nd ed. (London: Duckworth, 1987), and *Whose Justice? Which Rationality?* (Notre Dame, Ind.: University of Notre Dame Press, 1988). More generally, something of the turn to spatiality among several leading thinkers can be gained from an uneven set of essays collected in Mike Crang and Nigel Thrift, eds., *Thinking Space* (London: Routledge, 2000).

The claim that rationality "is always *situated* rationality" is made in the context of religious belief by Nicholas Wolterstorff in "Can Belief in God Be Rational?" in *Faith and Rationality: Reason and Belief in God,* ed. Alvin Plantinga and Nicholas Wolterstorff (Notre Dame, Ind.: University of Notre Dame Press, 1983), 135–186. One feminist insistence on "positioned rationality" is provided by Donna Haraway in her essay "Situated Knowledges: The Science Question in Feminism and the Privilege of Partial Perspective," reprinted in her *Simians, Cyborgs, and Women: The Reinvention of Nature* (London: Free Association Books, 1991), 183–201. Haraway characterizes the "view from nowhere" as the "God-trick"; a God's-eye view, however, might be better considered as "the view from everywhere."

Index